T0396846

# You Can Be the Next Einstein

# You Can Be the Next Einstein

George Jaroszkiewicz

World Scientific

NEW JERSEY • LONDON • SINGAPORE • BEIJING • SHANGHAI • HONG KONG • TAIPEI • CHENNAI • TOKYO

*Published by*

World Scientific Publishing Co. Pte. Ltd.
5 Toh Tuck Link, Singapore 596224
*USA office:* 27 Warren Street, Suite 401-402, Hackensack, NJ 07601
*UK office:* 57 Shelton Street, Covent Garden, London WC2H 9HE

**British Library Cataloguing-in-Publication Data**
A catalogue record for this book is available from the British Library.

**YOU CAN BE THE NEXT EINSTEIN**



ISBN 978-981-121-112-6
ISBN 978-981-121-209-3 (pbk)

For any available supplementary material, please visit
https://www.worldscientific.com/worldscibooks/10.1142/11575#t=suppl

Desk Editor: Nur Syarfeena Binte Mohd Fauzi

Typeset by Stallion Press
Email: enquiries@stallionpress.com

Printed in Singapore

To the memory of Gloria, James, and Mark

# Preface

*Every writer hopes or boldly assumes that his life is in some sense exemplary, that the particular will turn out to be universal.*

Martin Amis, [The Observer (1987)]

This book is aimed at students who may never have thought about doing science. Not just any old science at that, but a particular exotic branch of science called *mathematical physics*.

Before you put this book down as **not for you** (as *Sheldon Cooper* of *The Big Bang Theory* might say), give me a chance to entertain you with some stories about my experiences as a mathematical physicist. Perhaps those stories (or *anecdotes* as I call them) will convince you that, whatever your background (race, gender, culture, language, economic circumstances, religion, physical ability, ...), becoming a mathematical physicist is not necessarily a senseless ambition for you. I'll even encourage you if you've been taken in by the common delusion that you're no good at maths.

The more I write this book, the more it's beginning to look like my autobiography. I never intended that, but the fact of the matter is that I'm writing from my experiences, about circumstances that conditioned me and pushed me into mathematical physics. There is, in consequence, an autobiographical strand running throughout this book. Most of the various incidents and anecdotes described in this book are based on events that actually happened to me or I witnessed happening to other people.

All my anecdotes were chosen to illustrate some important points about becoming a mathematical physicist. Apart from illustrating those points, these anecdotes show that science is done by real humans in ways that are often humorous. My anecdotes are based on what I witnessed and I report them as I saw and remember them. Of course, my memory may be faulty on occasions. Others who were there might not have noticed what happened, or perhaps disagree with what went on (in which case they can write their own accounts in their own books). Be reassured that I always give as honest an account as I can remember about what happened, without distortion to sensationalize. I've not invented any of them. Those anecdotes are in shaded background, as are observations that I make at various points in the book (I like to throw in my two cent's worth of personal opinion occasionally).

All the people I write about existed, apart from *Sheldon Cooper*, and did the things I said they did, as I witnessed them. In order to create a bit of mystery, I'll refer to some of those individuals as *Our Resident Genius*, *Scruffy Boy*, and so on. Other people such as *Einstein* and *Stephen Hawking* will be referred to explicitly.

I start each chapter with a quote relevant to that chapter. One of the annoying things these days is the unreliability of information from the internet. I've tried hard to track down suitable quotes from verified sources that I possess in one form or another. I give those sources in the Bibliography. I want you the reader to have confidence that whatever I write in this book is genuine, particularly the anecdotes.

I decided not to play any *labelling* games in this book. By this I mean I'm not going to say this or that person was from this or that country, nation, culture, or ethnic group. Proper science doesn't concern itself with such issues. Sometimes you'll be able to work out a label for yourself. That will certainly be the case as far as gender is concerned. As far as I'm concerned, none of those things matter. Having said that, I am like you, the product of my own particular origins, past context, and life experiences. There's a chance I never knew about people important to you and your culture that should be mentioned in this book. Any such omissions are done in ignorance on my part.

The more I read about the history of science during the writing of this book, the more I got dismayed about the disadvantages and bias faced by women over the centuries. Gender inequality is one of the icebergs of society: most of it is out of sight. I discovered that there have been women whose contributions to science have been ignored or attributed to men. Although this book was written to encourage students from every background, I would like to believe that perhaps it will push more girls to go into mathematical physics.

I have to express my gratitude to quite a few people, such as my parents, who gave me all my opportunities and never questioned my decision to go into mathematical physics. I am profoundly indebted to the UK educational system that was in place during my childhood and later years. It gave me the opportunities that led to my career in mathematical physics. Along the way I had the fortune to be taught by inspiring teachers and lecturers and I shall always think of them as wonderful people. I will mention some of them in this book. Most of them have by now surely passed on in the *Great Scheme of Things*[1]. I regret now that I never told any of them how excellent they were. If you have any such teachers or lecturers, *tell them before it's too late.*

Fortunately, I still have members of my family around who have been so helpful, so I can tell them right now how grateful I am for all their support.

*George Jaroszkiewicz*

[1]I don't know what the *Great Scheme of Things* is precisely, but it involves life and death for sure.

# Foreword

I've structured this book as follows.

There are three *themes* in this book. The first theme, *Motivation*, runs from Chapter 1 to Chapter 12. It discusses reasons why mathematical physics may be a good career choice for you. The second theme, *Practicalities*, runs from Chapter 13 to Chapter 21. It discusses the business of starting up and working in mathematical physics. The third theme, *Technical Stuff*, runs from Chapter 22 to 24. It discusses various scientific paradigms such as classical mechanics, quantum mechanics, *String Theory*, *Many Worlds*, *Hidden Variables*, and notation, as well as possible directions of research if you do go into mathematical physics.

This book can be read serially, from Chapter 1, Chapter 2, and so on, but you won't lose anything if you read any chapter on its own, in any order.

*Anecdotes* are displayed in shaded background and are listed in the *Table of Contents* as subsections in boldface.

Various terms that you might not be familiar with, such as *postdoc*, *boondoggle*, and *realism*, are explained in Chapter 25, the *Glossary*.

Terms followed by a superscript are commented on in Chapter 26, *Notes*.

# Contents

## Chapter 1

# Introduction

*A new scientific truth does not triumph by convincing its opponents and making them see the light, but rather because its opponents eventually die, and a new generation grows up that is familiar with it.*

[Planck (1949)]

This book was written with young people in mind.

Perhaps I should rephrase that. This book was written for people with young minds. By that, I mean people who have not yet been fully conditioned to stop thinking for themselves. That will of course include some older people, but what *Max Planck*, the founder of *quantum mechanics*, wrote (quoted above) may be applicable here.

I'm not saying that this book is about a new scientific truth. In fact, this book is based on an old idea, *Nullius in Verba* (take no one's word for it), an idea that took a long time to become accepted by scientists as the correct way to do science. I'll have a lot more to say about *Nullius in Verba* throughout this book, particularly in Chapter 6, *Nullius in Verba*.

It's regrettable to tell you this, but there is evidence that some influential and senior scientists seem to have abandoned this principle. Two striking examples come to mind here. First, some *String* theorists appear to have abandoned *Nullius in Verba* in favour of an appeal to *mathematical beauty* (whatever that means) [Woit (2006)]. My second example is the unsuccessful attempt by believers in *Creationism* to have the *Royal Society* of London accept it as a scientifically valid concept (it is not) [Royal Society (2008)].

I expect a fair percentage of older scientists will object to what I've written above (and by implication, what Planck meant). It sounds wrong and unfair, but my experience over forty years has been that what Planck said is by and large true. Once they get established in their careers, people tend to be fixed in their ways of thinking (I don't exclude myself from this observation). They come to believe that the once-revolutionary ideas they had in their youth are truths and they no longer question their validity or value. An example of this is Einstein. In his younger days, Einstein developed revolutionary theories such as Special Relativity and General Relativity and even made significant contributions to the development of quantum mechanics. As he got older, however, Einstein retained a belief in realism, rejecting the non-realist interpretations of quantum mechanics that had become standard, such as the *Copenhagen interpretation* [Isaacson (2007)].

It should always be kept in mind that *Nullius in Verba* applies just as much to one's own views as to the views of others.

As for myself, although I've always applied *Nullius in Verba* intuitively all my working life, I became aware of its universal applicability[1] only in the last five years or so, when I started to write my books. I found it applicable not just to my work but in my normal life. This book is aimed at people who could benefit from knowing this principle and practising it. By and large, that means young people, mainly not over thirty.

## 1.1 Conditioning

The reality of human life is that everybody gets conditioned, sooner or later. That's what upbringing, experience, and education, are all about. We get conditioned to be respectable citizens or hardened criminals, to believe in some religion or none at all, to believe our country is best or that it needs to change, to take exercise or to over-eat, and so on. Societies can't function properly without some sort of conditioning of their members. Even anarchists and revolutionaries have been conditioned to be anarchists and revolutionaries, although they don't know it. I have been conditioned by my parents, friends, schools, universities, life experiences, and so on. But I have also been conditioned by my scientific training to understand that I have been conditioned. That understanding has often helped me to see things for what they are, or in some cases, are not.

Why is this important? If you stop and think about it, you'll agree, I hope, that the way we humans *think* is all we've got that makes us human. We're pretty pathetic as physical creatures. We have to wear clothes, we take ages to grow up, and we treat each other pretty badly a lot of the time. It's only the way we think that gives us an edge in a potentially hostile universe[2].

It's an empirical fact that the *way* we think, and not just *what* we think, is not fixed until we're into our twenties, and even then we can, literally, "change our minds" about beliefs that we once held to be absolute truths [Giang (2015)]. When I refer to the *way* we think, I'm not referring to neurones, the hippocampus, prefrontal lobes, and all of that stuff. There's not much you can change there by yourself, except in a bad way. What I mean is that you can alter your beliefs and interests by adopting new habits. By and large, the more you repeat an action, the firmer your brain gets rewired in response. For example, every time I get into a car, I put a seat belt on. I would feel uncomfortable otherwise. I don't think millions of years of evolution has structured my DNA to put that precise thought in my brain. Until cars and seat belts were invented, no one on this planet would even have thought about such an idea. The fact is, I acquired that habit only when the United Kingdom Parliament passed a law about thirty five or so years ago. At first, I put my seat belt on every time because I feared getting a fine from the police. After a few months, I did it automatically, because I had become conditioned to do it. It had become normal.

Do you like vegetables? I don't and tend to avoid them. I know that's an unhealthy tendency, but I used to believe that there was little I could do about it. Fortunately, research has shown that it is possible to condition yourself to like them [Heathers and Nickle (2019)]. If I've ever gone to sleep without brushing my teeth (which sometimes happens when I'm travelling), I feel uncomfortable. That's been conditioned into me since childhood.

It's an important fact that people tend to be quite resistant to change in their beliefs. If people were easily manipulated into changing what they thought was normal, then societies would be more unstable than they usually are. Social conditioning has its own form of inertia in this respect. For example, fox hunting was banned by law in Britain in 2004. Even now, there is a strong lobby to rescind that law. As Planck suggests above, some people never get used to the new ways.

I often look for *analogies* when discussing complex issues, because a decent analogy can help me see those issues from new perspectives. In this respect, we can think of conditioning as analogous to writing a book. The individual neurones in our brains are analogous to letters. Individually, letters tell us nothing about the book as a whole. We connect letters together to create words and these represent elementary concepts such as names, adverbs, and adjectives. But words by themselves don't tell us much either. Words are connected into sentences, and these can express complex ideas. Even then, sentences by themselves don't tell us the whole story. It is only when we have the whole collection of sentences, organized into paragraphs, sections, and chapters, do we arrive at the meaning of the book itself.

A fundamental point about books is that, whilst any two books in a given language will be constructed out of identical letters and identical rules of grammar, those books could be about totally different topics having nothing in common, or even about the same topic and contradicting each other. So it is with conditioning. One person may be conditioned to believe in a God whilst their friend does not, but both have the same sorts of neurones in their brains[3].

When you train or do exercises, you are effectively wiring up the neurones in your brain into patterns that will end up conditioning you into a certain modes of thinking. It's not easy and takes time, usually a lot of time, measured in years. That's why you go to school for years and years.

Sometimes, conditioning can be dangerous. For example, visiting certain web-sites regularly can influence the way we think and, in some cases, act. Ever heard of online gambling?

When I refer to conditioning, I mean an acquired habit that through practice becomes more and more part of your psychological make-up and contributes to all those things that define you as a person. The problem is that you may not be aware this process is happening to you. There's scientific evidence that people's neural pathways are still being connected up to about the age of twenty five or so. I don't think it's a coincidence that people start their doctorates around about that age (plus or minus five years).

Conditioning takes many forms. Brushing your teeth every day at seven o'clock in the morning for one month may condition you to doing it at that time for the rest of your life. Going to church, synagogue, mosque, or temple regularly when you're a child may condition you to go there for the rest of your life, or put you off ever going again when you're an adult.

Meeting an unpleasant person just once may condition you to avoid them thereafter. By practising a musical instrument or a sport an hour a day, you become better and better at it.

## 1.2 Reconditioning

Why am I going on about conditioning, when the title of this book is *You Can Be the Next Einstein*? The way I see it, conditioning is *probably* what made Einstein who he was. I say *probably* because there was a suggestion, even when he was alive, that his brain was somehow different to "normal" brains. That suggestion is relevant to us in this book because, if it were the case that his brain was structurally different to that of other people, then this book would be pointless. I don't want to give too much away here [spoiler alert], but the evidence suggests that it was not that Einstein's brain was particularly different to other brains, but that it was wired up in a special way, and it was that wiring that made him the scientist that he became. I'll discuss this topic further in Chapter 11, *Einstein's Brain.*

My suggestion is that, given that such rewiring takes place all the time in our brains by processes such as *experience* and *education*, there is hope for us all. I believe that if you take the right steps, you *can* condition yourself towards becoming a mathematical physicist. Of course, it involves hard work, but that's the case with any worthwhile ambition.

In my experience with students, the critical factor that can often overcome lack of natural talent is *motivation.* If you have strong enough motivation, you can endure the hard work required to do well in many disciplines. Without motivation, however, great talent can be wasted. I would not advise anyone to go into any great enterprise or career if they were not motivated properly to do so. Don't start new activities in your life just because you have to or it's expected. Try to find positive reasons instead.

**Not Inspiring**

This first anecdote of mine isn't humorous. It wasn't amusing at the time and it still makes me shudder.

I don't know about you, but certain professions such as medicine and law always seemed the ones to go for as a career. First of all, they're usually well paid. Besides that, there's always a need for doctors and lawyers, so as professions, they seem reasonably secure.

By about age fourteen or so, I had made up my mind I wanted to be a scientist. But there was the nagging thought that medicine might be a suitable alternative. After all, medicine is a branch of science these days. I'm referring to your standard medical discipline of course, involving sophisticated equipment and scientific trials to establish the validity of treatments, not holistic hokum based on pseudo-scientific intuition. So I was quite responsive when our teacher took us to see a short film about surgery. It was part of the school's efforts to guide students towards various careers, such as medicine.

I can see it all vividly in my mind even now. My class took a coach to the venue, which was actually a disused skating rink on the edge of town. We went in and I sat at the back of the balcony, eagerly anticipating an interesting film about surgery. Why not?

And that's when the nightmare started. The film was set in an operating theatre, in rather garish colour with lots of reds and greens. That wasn't the worst thing though. The subject was *proctological surgery*. I'm not going to explain what that means. You can look it up for yourself. All I can say is that, as a potential topic for recruiting surgeons, it did the opposite to me.

But the worst was to come. The film started with a short introduction, and then the operation actually started. With one incision of the surgeon's knife, never-ending streams of dark red blood started to ooze down the screen, and kept on coming. Think of Niagara Falls, but with blood.

After that film, I knew for sure I never wanted to go into medicine.

There is a positive side to this story. Although that film conditioned me against ever going into medicine, I developed an immense respect for those hardy individuals who do go into medicine and can deal with the inevitable unpleasant aspects of it, such as the mess you get. If that film had been visually gentler, I would not now shudder at the mere mention of any injury.

The moral of this story should not be lost on teachers, I hope. If you want to inspire your students towards any subject, do make it attractive, not repulsive. It's no coincidence how often we hear some famous person say how inspired they were by some kind teacher to go into their life's career, and how often people identify a dislike of some subject such as mathematics with an unpleasant teacher.

## 1.3 Taking the plunge

This anecdote may relate to your experiences. It's about making decisions that could change your life in a serious way.

### The Spinning Floor

There I was, aged fifteen, three months into my Advanced Level Chemistry course and starting a December class exam. Advanced Levels took two years to do, normally, and then it would be on to University. In those days, a student would take three Advanced Level subjects, invariably biased firmly towards maths and science, or else biased towards languages, humanities, and the arts. That meant that by age sixteen, we were all irreversibly straightjacketed into one side of the art-science divide or the other. By that age, it was assumed people knew whether they were going to be Science based or Humanities (Arts) based in their future lives.

By age fourteen I knew I wanted to be a scientist. It was only much later in life that I realized that, for many people, that would not be the case for them. My choices in the September of the year in question had been Physics, Pure Mathematics, and Chemistry as my three Advanced Level subjects.

The problem that emerged for me by the December of that year was that I had come to the dreadful realization that I could not stand Chemistry. The subject bored me senseless. I thought it required no thinking about the principles of physics, but was more like doing a gigantic jigsaw. You had ninety two kinds of pieces (the elements) and you just made pictures (molecules) with them, without any obvious end to the possibilities.

Interestingly enough, I've recently heard of a suggestion from a well-known and respected chemist that the subject is now more or less complete [Bulkin (2019)], in the sense that "all" that remains to be done is to apply well-known principles in new situations, such as nanotechnology (more-or-less the art of making atomic-size machines). So perhaps I was ahead of my time. I was always clear in my own mind that there is a fundamental difference between *discovering* the laws of science on one hand and *applying* them on the other.

I think that difference is what's meant by "R & D" (Research and Development). The side of mathematical physics that excites me most is the research end of things, *discovering* how the Universe works, rather than the development side, which is applying that knowledge to specific situations.

Back to the story. There I was, one cold, dark, winter's afternoon, sitting at a desk in a classroom, starting a Chemistry "Mock" exam. And that's when **the floor started to spin**.

No, I was not on any drugs. I have never ever taken any illegal substances (honestly), as I have always worried about how they might affect my brain. What happened next was one of the strangest experiences of my life. I suddenly felt as if any ambition I had in me had just packed its bags and left. All initiative too had deserted me. I lost the will to live. I wanted to tear my eyes out. Time seemed to get slower and slower for me, whilst the floor and the rest of the Universe started to spin faster and faster under my chair. I was doing something I hated and simply could not endure it or stop it. I was powerless to get out of my situation. Have you ever felt like that?

I did finish that Chemistry exam, but I don't remember whether I passed it or not. That question became moot, as *Sheldon Cooper* would say, because I decided there and then *to abandon ship and do Applied Mathematics and not Chemistry.*

As soon as I could, I spoke to our Mathematics teacher and convinced him to let me jump into the Applied Mathematics class, three months into the two-year course. I succeeded and that jump turned out to be the second most important change in my life. One of its long-term consequences, for example, is the writing of this book, more than five decades later.

Starting three months in was not hard at all. I just loved the subject. It introduced me properly to Newtonian mechanics. Don't assume Newtonian mechanics is old hat. It's a truly fantastic discipline. It can be applied to about ninety nine point nine percent of the world you think you're living in. You have to have sophisticated technology to see where it fails. When you drive a car fast around a corner, a little appreciation of the forces involved can save your life. If you ever crawl over a sloping roof to fix a broken roof tile, an understanding of friction, normal reaction, and gravity, could save your life. If you know something about *inertia* (the tendency of objects to keep on moving), then you will not feature on any of those video clips that you see, where some idiot cuts a tree down and it crashes onto his house. Or a proud father-of-the-bride tries to jump over his daughter's wedding cake and crashes right into it. Almost all of those clips demonstrate some aspects of Newtonian mechanics.

I'm in two minds about the above anecdote. You see, whenever a student from another department came to me in the maths department where I worked, asking to transfer into Mathematics from some other degree such as Physics, I always asked them *why?* If they said that they had failed exams in their original degree and needed to change to a subject they might pass in, I would advise them to find a better reason. *Don't change from course A to course B just because you've failed exam A. Find positive reasons for changing to Subject B.*

So in my case, did I change from Chemistry to Applied Maths because I failed Chemistry?

I don't think so. First of all, I don't recall whether I failed or not. The fact is, I developed a loathing of Chemistry *before* I sat that exam. My real failure was in choosing Chemistry in the first place. I had not properly thought out my true interests before making my three Advanced Level choices.

Fortunately, I was able to correct that initial wrong-for-me choice. So my advice is: think carefully about your future life and plan accordingly. If you do come to a critical decision, *act on it as soon as you can, even if it looks a bit late.* **But do be sure it's what you really want**.

## 1.4 Change yourself

One of the fundamental aspects of conditioning is that it obeys no rules of morality. That's what makes it potentially dangerous and potentially beneficial. It's a fact that regimes of all sorts of work hard to condition their citizens into conditioned patterns that suit those regimes. Some are good, some are manifestly bad. I think it's important for a person's self-development to be aware of this. For example, Einstein was well known for having a strong aversion to the militaristic form of conditioning prevalent in the society in which he lived in his youth [Isaacson (2007)].

Most conditioning is imposed on us by laws, local traditions, national outlook, and educational systems. If you want to become a mathematical physicist, you will have to do it yourself, by choosing to recondition your own patterns of thought. Be aware that such changes do not come without some cost. The most significant cost here is some change in your personality. You will not be quite the same person afterwards compared to the person you were before. The important point is, though, that at least *you* will be the one deciding to make the change, not other people or organizations.

I can reassure you that you will not become a worse person if you become a mathematical physicist, unless you choose to be. Such a choice is discussed in Chapter 7, *The Gates of Hell.*

# Chapter 2

# Role Models

*Example is not the main thing. It is the only thing*

Albert Schweitzer [Schweitzer (1952)]

*You Can Be the Next Einstein.*

What a peculiar title for a book. On the face of it, wanting to be someone else sounds like a pretty bizarre ambition. How can anyone want to be Einstein?

Of course, I don't mean it literally. I mean, you may want to do marvellous things in theoretical science just like Albert Einstein did. So really, it means using Einstein and his accomplishments as an example to follow. A *role model*, in other words.

It's a fact that throughout history, people have had *role models*. That means having some well-known person or historical figure to look up to. Someone who achieved great things, and is an example to others. Alexander the Great[1] was himself a role model for a number of soldiers, such as Julius Caesar. Quite late in his life, Julius lamented the fact that Alexander had achieved more at the same age as himself. History records, however, that Alexander carried around with him stories about Hercules, a hero in Greek mythology. So Alexander had Hercules as *his* role model.

When we talk about role models, we really mean their achievements. If our role model is say a great sportsman or sportswoman, we imagine ourselves winning the competitions that they won, such as in the Olympics. When we have a role model of a singer, we imagine emulating their successes in music.

So it is with Einstein. He is generally thought of as the greatest scientist of the Twentieth Century. Certainly his name crops up quite frequently when non-scientists talk about smart people. He had a hand in the creation of two great scientific theories, General Relativity and quantum mechanics[2]. I'll have more to say about those subjects later on in this book. Certainly, Einstein was not exactly alone in the development of those theories. Moreover, some of his views about quantum mechanics (which he ended up not liking at all) are not nowadays mainstream. But throughout the first fifty years of the Twentieth Century, what he thought about General Relativity and quantum mechanics made other people take notice. He is a *great* role model of a theoretical scientist.

By the way, I've got to be careful here in several crucial respects.

## 2.1 Einstein was human

On the face of it, this book puts Einstein on a pedestal. Yes, but only as a mathematical physicist. Although he is often portrayed as a great humanist, he had his personal flaws. If you look hard enough, you will find out about them. They existed because no one is perfect[3].

## 2.2 Gender

I've mentioned men up to now but there have been brilliant women who are wonderful scientific role models. You may have heard of Marie Curie, who won **two** Nobel prizes in science. Perhaps you've heard of Mary Somerville, Ada Lovelace, Emmy Noether, or Rosalind Franklin. They were brilliant mathematicians and scientists. There are many more great women in the history of science, such as *Hypatia*, if you look.

There are brilliant women mathematical physicists active right now. If you're a girl, you can aspire to join them. Mathematical physics is not biased towards one gender in any way. It's the case that in English, the language in which I think, speak, and write, the term *mathematical physicist* has no gender, meaning it's not identified as male or female[4]. Now it may be the case that in your language (assuming your main language is not English), the expression *mathematical physicist* is gendered, but I can reassure you that that would be down to linguistic convention and not related to job description. There is no reason to take word gender as anything more than convention. For example, the French word for *person* is *personne*, which is grammatically feminine, even when applied to males.

I believe that the historical preponderance of numbers of men over women in science is down to the severe cultural pressures of educational and economic bias against women in the past, a bias that has not yet been eliminated properly in many places around the globe. Fortunately, the growing proportion of women going into science and mathematics at university level suggests that there may well be a reversal in the historical statistics in the future. I look forward to the time when there will be no need to mention gender in a book such as the one I am writing. The following anecdote tells me that such a time has not yet come.

## I was told

Over the years during my time in the mathematics department where I spent most of my career, I had a variety of administrative jobs. Everyone got them. The incident I'm relating was when I was interviewing candidates for admission to the department to do a mathematics degree.

On this particular occasion, a fine young girl came into my office for a ten minute interview. I had received her University Application documentation and it was impressive. She came from an all-girls school, had already passed with outstanding grades at the "Ordinary" level, and was predicted to do spectacularly well at the "Advanced" level. By any yardstick, she was just the sort of entrant we were looking for.

My favoured interviewing technique was not to ask technical questions. We were supposed to, but I never really understood the sense of that. What's the point of asking a nervous young person sitting in front of you how to integrate cosine squared of $x$ from $x$ equals zero to $x$ equals $\pi$? The exams would take care of all that technical side of their abilities.

I always thought it much better to ask about the applicant's ambitions, how they saw mathematics, or in many cases, mathematical physics. It's much better to understand *motivation* than *ability*, because in my experience, there's nothing sadder than a technically brilliant student with no motivation. I have found over the years that you cannot give ambition to a disinterested student. The best you can (and should do) is to provide an example (be a role model, essentially), and hope that it rubs off onto them. They have to find ambition for themselves, and sometimes, none ever comes. Parents and teachers who push their children into subjects unsuitable for them are risking life-long regrettable consequences.

As we chatted, I glanced over the girl's *Personal Statement*. That was a part of the application form where the candidate writes about themselves and what they were looking for in life. What caught my eye was that she had written that she wanted to be a *school teacher*.

Now really don't misunderstand me here. There's nothing wrong about wanting to be a school teacher. It's a great profession that deserves the highest respect. But I was intrigued. I could see from her achieved grades and predicted grades that she was truly outstanding. So I asked her: "*What inspired you to want to be a school teacher?*"

"*My Careers Mistress at school* **told** *me to write that in my Statement.*"

I remember how dismayed I was after that interview. Was *that* the only Careers guidance given to such brilliant girls in those days?

Fortunately, not many years later, I would be interviewing young women applying to do mathematical physics who had already decided for themselves what they wanted to do in life. Yes, some of them wanted to be school teachers, but that would have been *their* informed choice, not foisted onto them by the conditioned social expectations of older people. Others would go on to do doctorates in mathematical physics.

I hope that you will see from this that this book is aimed at encouraging boys *and* girls into mathematical physics.

## 2.3 Race, ethnicity, and nationality

Great role models in science are certainly not confined to white Europeans, either. Einstein himself, of course, was born in Germany and had Jewish ancestry. The list of mathematical physicists from other continents is long and impressive. Three such names come to my mind immediately: the particle theorist *Mohammad Abdus Salam*, the astrophysicist *Subrahmanyan Chandrasekhar*, and the particle theorist *Shinichiro Tomonaga*. I know those names because my own interests in astrophysics and quantum field theory led me to encounter their works. But as with the women scientists, you can easily find many more examples if you but look.

Here is an example that shows how science is not confined to any particular culture, race, continent, or age.

## Mpemba's Ice Cream

There I was, quite a few years ago, sitting with my maths department colleagues one lunchtime in the University Staff Room, drinking coffee, making small talk, and reading that illustrious magazine *New Scientist*.

The reader will be forgiven for thinking that I spent all my time doing that sort of thing. The fact is, talking to people and reading articles is part and parcel of doing science.

I turned the pages of the magazine and came across a curious and interesting story. It concerned a young schoolboy named *Erasto Mpemba* in Tanzania, who noticed that when he attempted to cool an ice cream mixture, *it cooled down faster from relatively hot temperatures than from relatively cooler temperatures.*

That was an extraordinary observation. It's contrary to conventional physics wisdom and expectation, according to the following line of thought. First, suppose you prepare your ice cream mixture at fifty degrees centigrade and it takes half an hour to cool down to room temperature. Now prepare another sample, but this time start it at eighty degrees centigrade and time its drop in temperature. Say it take ten minutes to drop from eighty degrees to fifty degrees. At that point, it should now take another half hour to cool down to room temperature, just like your first sample. So going from eighty degrees to room temperature should take ten minutes plus thirty minutes equals forty minutes, which is longer than the time taken by the first sample.

Erasto found that it was sometimes the other way around. Sometimes, a sample at a relatively high starting temperature cooled down *faster* than a sample starting at a lower temperature.

Erasto asked a visiting physicist about what he, Erasto, had observed. To his credit, the physicist took it seriously, although at first, other pupils and Erasto's teachers laughed at the very idea. Remarkably, the physicist and Erasto eventually published a science paper on the phenomenon [Mpemba and Osborne (1969)] and it has been of interest ever since.

There has been some debate about whether the *Mpemba effect* (as it's often referred to) is real, and the matter remains open to further investigation. For example, Burridge and Linden discuss the cooling of hot *water* [Burridge and Linden (2016)] and conclude that there is no evidence for the Mpemba effect, *in water*.

Yes, I agree with that. But strictly speaking, Burridge and Linden's results are irrelevant, because Mpemba's original observations were not about water but ice cream mixture, something significantly different.

I think that there may well be *structural* effects going on with ice cream mixture. What I mean is this. Say you had a really hot starting ice cream mixture. At that temperature, it would be bubbling and hot gas would be flowing through to the surface as a result. As that sample cooled, some of those bubbles might solidify or gel, creating sponge-like tunnels in the structure of the substance. This would give a different internal architecture compared with that of a perfectly homogeneous mixture. If true, this would mean that starting from say eighty degrees, a sample that had cooled down to fifty degrees would be structurally different to a sample that had never been at the higher temperature but started at fifty degrees. I can only speculate on this as I'm not a materials scientist, but I think that the Mpemba effect may demonstrate the critical importance of *context*.

By context, I mean the taking into account of *all* the relevant circumstances under which something is done. If you ever get to do any quantum mechanics, you will run right into that concept. Context is one of those concepts that people tend to forget, but in my experience, context helps explain a lot that otherwise looks mysterious and baffling.

The point about this anecdote is that science is international. It is for everyone and can be done by anyone, if they so choose.

Some role models are well known because of the way we are influenced these days by social media such as television, film, and the Internet. You must have heard about Stephen Hawking. As a role model for disadvantaged people, he is iconic. He had the most incredibly difficult medical condition to live with, *motor neurone disease*. That forced him to move around in a wheelchair all his working life. Yet he conjured up from his mind deep and insightful theories about black holes and the origin of the universe. What a great role model he was.

Here's a real-life incident I witnessed involving Stephen Hawking.

### Interruption

As a theorist, for many years I used to go every December to a two-and-a-half day meeting of UK theorists at the *Rutherford Laboratory* at *Harwell* near Oxford. The idea was a series of hour-long talks about the latest hot-shot developments in theoretical and experimental particle physics and cosmology over the last year. These meetings were invariably packed out and the lectures were usually incredibly interesting (but not always).

That particular year, one of the speakers was Stephen Hawking. I can't remember if he was the first speaker, but regardless, he was *Top of the Bill*, as usual. Always popular, he was a brilliant speaker, with his mechanical voice synthesizer exactly as you may have heard on *The Big Bang Theory*. Well before he started, the auditorium was packed out. I myself was seated right at the back. That's always been my favourite place in lectures. You can see everything that happens. I'll have some more to say about *sitting at the back* later on in this book.

It was dramatic. The stage was empty, the lights were low, the audience buzzed with anticipation. Then, at the appointed time, the side door to the stage opened mysteriously, and then Stephen Hawking came in by himself on his motorized wheelchair and the door closed behind him. As he glided in, he looked like some visitor from a distant planet, come to Earth to give us mere mortals deep knowledge. I can't remember quite what his topic was, but it would have been about gravity, black holes, the beginning of the Universe, and all that stuff. No joke, this was *The Big Bang Theory* in real life.

We all sat there entranced. He had a way of communicating that was direct and had humour. He was entertaining and authoritative at the same time.

He went on for about a quarter an hour. Then ... it happened.

There was a sudden noise from the side door from which Stephen had entered the stage. Stephen stopped. We all looked at the source of the noise. The door opened slowly, and there then stepped through it onto the stage a tall, imposing man in uniform. A security guard.

He stood there, framed by the open door behind him. A tall man with authority. A man in a military style cap and a smart uniform. Like a sergeant major inspecting his recruits, he looked up at us and scanned us from left to right. The assembled horde of several hundred theorists, experimentalists, professors, and students, brought from all over the UK and many other countries at great expense, stared back at him in baffled silence. Stephen had turned and was also looking at this interloper. What was going on?

After what seemed like an eternity of silence, looking up at the assembled multitude, the security guard spoke these mighty words:

"*Will the owner of a white Ford Cortina please come and switch off their car lights in the car park.*"

I've got another anecdote about the same meeting, but in a different year, which I'll relate in Chapter 17. If anything, *that* anecdote gives a good message about seizing your opportunities when they arise. *Carpe diem* (Seize the day), as the Romans used to say. That's a really good bit of advice.

A role model you may be familiar with is *Sheldon Cooper*. Of course he doesn't exist, he's just a character in a long-running comedy series known as *The Big Bang Theory*. Nevertheless, his fictional ambitions and achievements will surely inspire some of my readers to become theorists themselves. What you will have noticed about Sheldon Cooper is that he's portrayed (brilliantly, I may add, by Jim Parsons ) as having a number of quirks or peculiarities of character that reinforce the stereotype image of scientists as having something wrong with them. Sheldon is a control-freak and lives in a nerd-world of computer games, science-fiction, and yet does great science.

What can I say about that? Only this. There is no logical or proven connection between being a scientist and doing any of the other stuff that Sheldon does. I've come across many scientists who are brilliant and yet absolutely "normal" in any sense of the word. But then again, I've also met great scientists who are really weird.

Einstein himself was well-known for some eccentricities. Apparently, no one ever bought him a comb or introduced him to a good barber, at least, so it appeared later on in his life. Apparently, he was not fond of wearing socks. That's not so unusual in many parts of the world, and it saves a bit on laundry.

## 2.4 The limitations of role models

Readers who are familiar with *The Big Bang Theory* series will know that Sheldon Cooper starts in Series One doing *String Theory*. He is passionate about it and mocks theorists such as Leslie Winkle who is engaged in *Loop Quantum Gravity*, regarded as a competitor theory. The writers of the program were obviously well advised by their scientific advisors, because eventually, Sheldon loses faith in *String Theory* and goes on to win a Nobel Prize with Amy Farrah Fowler in another area of theory.

This touches on an important issue that anyone going into mathematical physics should take into account. Having a role model is often useful, but in the long run, of limited value. Eventually, you have to start thinking for yourself and do your own thing in your own way. If you follow any role model precisely, then you'll just be repeating what they did. That's the opposite of what science is all about. Do your own thing diligently and properly, and then perhaps *you* may become a scientific role model for others.

## 2.5 The song, not the singer

The land of Academia is a great and wonderful land, full of opportunities and marvellous people. It has many fine Universities and Colleges, vying for your attention.

It also has back-waters and dead-ends.

Suppose you have applied to some fine Universities, but none want you. Accordingly, you have resigned yourself to doing a degree at a little-known University, with no international prestige worth mentioning. Are you doomed to less of a glorious career than your friend, who got into a prestigious University?

Here's where your character, your personality, matters. In mathematical physics, no famous University or College will give you a great career just because you went there. You have to make your career yourself. If you have the determination and the motivation to do a doctorate in a little-known department where there are no internationally acclaimed experts and your supervisor is disinterested, *do not be discouraged.* Having a disinterested supervisor can happen anywhere. What matters is *you.* If you do good work, that's all anyone can do. No one will give you acclaim for being Einstein's last collaborator or Hawking's last student. They will ask you what *he* was like, not what paper *you* just wrote. Being associated with a famous mathematical physicist can in fact be a great let-down.

In my view, the important point to focus on, should you apply to do a doctorate in mathematical physics, is the subject, not the supervisor. If you go for the place, or the supervisor, then be careful. None of those things will give you motivation. You have to discover that for yourself.

I will agree that being in a well-funded, active, internationally acclaimed department can be an advantage. It gives you a smug and shallow feeling of superiority when you tell people that. Being in a low-key, badly funded and low-morale department can be depressing. If you find yourself in such a place, my best advice is that you should focus only on the discipline and that will get you through. No one really cares where Einstein went to school or college, except in books about the history of science. What everyone is interested in are his theories, which he developed by himself in various places. One of them was a Patent office in Switzerland.

# Chapter 3

# Is This You?

*...stupidity is much the same all the world over; a stupid person's notions and feelings will be simply the ones that are prevalent in the social circles he or she moves in.*

John Stuart Mill, [Mill (1869)]

This book is intended for the unusual young person. Perhaps it's you.

No. I don't mean *abnormal*, or suchlike. I really don't know what "abnormal" means, other than as a term of abuse. I don't even mean *physically different* to people around you, such as having tattoos all over your body, or having one arm, or anything like that. I don't mean having a different skin colour, family background, religion, or language, compared to people around you, either. All of those differences I would say are perfectly normal but individual to you in your own particular way. There's nothing wrong with any of that.

What I mean by *unusual* is this. Do you play computer games, or go on social media (Facebook, Twitter, Instagram, and other displacement activities), or watch *The Big Bang theory*, *X-Factor*, *Pop Idol*, *Netflix*, and so on, yet you still think there is something missing in your life? That's what I mean by *unusual*.

Perhaps you're not in such a position. Perhaps you do all of those things and think they're absolutely wonderful and fulfilling. If that's your choice, fine. On the other hand, suppose you don't have any computer or mobile phone (it happens) and you work hard all day at a job in a brick factory for a pittance. Yet, despite that, you still manage to think about a greater world out there, beyond your immediate experience, waiting for you to explore it. Not a world of wealth and privilege but a world of great thought and imagination, and you would like to see it.

Perhaps you do have such thoughts, but you haven't let on to any of your friends that you think like that, because you are worried they would laugh at you, or think you were weird. So you keep your feelings to yourself.

I don't know the exact statistics, so can only guess. Out of say a thousand teenagers, one or two may feel as if life has more to offer than endless video games, smart phones, sports, cinema, music, social media, and so on. That could be you. You may well be one of the unusual people I mean. Then do read on. I've written this book for you.

## Profession

There is a science-fiction story called *Profession* by *Isaac Asimov* that touches upon the theme of this chapter [Asimov (1957)]. That story is regarded in some quarters as an indictment of those contemporary educational systems around the world that focus extensively on examination performance and little on creativity. You can find this story online.

The hero of the story is a young man called *George Platen*. He lives in a so-called *House for the Feeble-Minded*, an institution that looks after apparent misfits who have been rejected from society because their brains failed certain tests. In the future in which the story is set, humanity has colonized other planets. Every year, people who have been conditioned to pass certain technical examinations are sent from Earth to run those planets.

But not people like George. It looks as if he is a total failure.

The problem is, George doesn't think so. So he escapes from his *House for the Feeble-Minded* and tries to get to another planet, where he can work normally like everyone else.

To cut a long story short, George fails to escape, but ends up discovering that in fact, he is not feeble-minded at all but special. Indeed, he finds out that the correct name for his *House for the Feeble-Minded* is really an *Institute of Higher Studies*, and it is there to help him in his eventual career. For it is George and rare individuals like him who create all the new technologies that the "normal" people use on the newly colonized planets.

As Asimov writes on the penultimate page: "*Somewhere there must be men and women with capacity for original thought*".

I don't think it should matter from which background you come. You may be male or female. You may be from a long-established rich family or you may be from an impoverished first or second generation immigrant family. Your family may be devoutly religious or agnostic. Perhaps you have an uncle who is a prison governor or a prisoner on a drugs offence. Perhaps your family lives in a mobile home or in a palace surrounded by high security fences and guards. Perhaps your parents are academics or no one in your family has ever gone to university or even passed an exam. Perhaps you are in a refugee camp. None of these factors should not, and need not, matter, if you set your mind to becoming a mathematical physicist.

It's possible you've never heard of people called mathematical physicists. Think again. Surely you've heard of Isaac Newton, Albert Einstein, and Stephen Hawking. They were all mathematical physicists. If you have ever watched that hit television series *The Big Bang Theory* you will know all about *Sheldon Cooper*. He's a mathematical physicist. If there's anything about those individuals that makes you think about being like them, this book may give you a pointer or two towards that ambition. I give more details, pointers, and guidelines in later chapters about that particular term *mathematical physicist.* I hope to convince you that being a mathematical physicist is just as good a career as being in say a pop group, or being an accountant, or even a banker. That's one of my main objectives in this book.

## 3.1 Do or do not. There is no try

I cannot promise miracles. I cannot say that you *will for sure* achieve such an ambition. I can give no guarantees, because *life is complicated.* Many talented people find themselves in circumstances that work against them developing those talents. For example, the sons and daughters of farmers may be required to follow their parent's footsteps and work on the farm.

In such cases, dreams and ambitions often get thwarted by circumstances apparently beyond a person's control. However, the determined individual will attempt to circumvent their difficulties. The golden rule is: *if you do not have a go, then you certainly will not succeed.* It's up to you whether you reach for the stars or stand there and watch them.

## 3.2 The journey, not the destination

An important factor that has helped me come to terms with my own life experiences was the realization that success is not an end in itself. It's the journey towards success that is the fun. It's the hard work, the daily grind, the toil and trouble, the friends that we make and meet on the way, that gives value and meaning to our lives, not the prize at the end.

You may disagree. In the world as it's currently structured socially, it's generally assumed and constantly drummed into us that only success counts, that no one remembers the second-placed. It's win, win, win at all costs. I think that's shallow. Certainly, you may get your prize, your medal, your reward. But how long will that moment of glory last? Today you may win the trophy, but next year, it will be someone else. Even if you do win a Nobel prize, how long are you going to look at it? How does having several billion dollars in the bank beat having real friends? I say, *think* a bit before you buy into the *win at all costs* conditioning. There's more to life than winning.

I will admit that there is one good argument in favour of the winner mentality: *if you don't try to win, then that's a recipe for stagnation, for mediocrity.* However, that statement needs qualifying: *win for the right reasons.* Don't win to beat the other guy, but to beat the person you are. Improve yourself, don't defeat others. If you aim to do mathematical physics because you want to get a Nobel Prize, then go see a psychiatrist. But if you want to do mathematical physics because it will enhance your appreciation of this wonderful Universe, then *that's* what I'm talking about. The mathematical physicists who developed the theories upon which our civilization is based were driven by intense curiosity and a desire to understand the world they were in, not by commercial success.

*You Can Be The Next Einstein*? Nonsense. I chose the title of this book just to catch your attention. I should have entitled it **Don't try to be Einstein, be yourself**. Try to achieve your ambitions on your own terms, as *you*, not as a clone of Einstein, Madame Curie, Stephen Hawking, or even Sheldon Cooper.

# Chapter 4

# Mathematical Physics as a Career

*We have reached the point, it is painful to recognize, where the only persons accounted wise are those who can reduce the pursuit of wisdom to a profitable traffic.*

Giovanni Pico della Mirandola, [della Mirandola (1496)]

Before I go on further in this book, there's a few questions I should address about mathematical physics as a career.

## 4.1 Does mathematical physics pay as a career?

Perhaps the most obvious practical question to ask about mathematical physics is whether you could survive financially being a mathematical physicist. My experience is that mathematical physics has given me an immensely satisfying career that gave me an income level that suited me. As a career, it's at the income level of university professionals, ranging from graduate students to professors. You will not starve as a mathematical physicist.

## 4.2 Does mathematical physics give you versatility?

One of the great strengths of a training in mathematical physics is that it gives you confidence in mathematics *and* in physics. That should give you an enormous range of possible career choices, should you decide to go out into the wide world beyond Academia after your first degree in mathematical physics. I noticed on a number of occasions over the years that our top mathematical physics student outperformed in their final mathematics exams (as measured on percentage scales) the top single Honours mathematics student and outperformed in their final physics exams the top Single Honours Physics student.

In some chapters, I give some anecdotes about my experiences looking for and applying for jobs. I can assure you that my training in mathematical physics gave me the versatility to apply for, and get, jobs which appeared to require specialist experience that I did not have at the time. What I found was that the academic employers concerned realized that my training in mathematical physics gave me the versatility to fit into their job requirements within a relatively short time.

## 4.3 Is mathematical physics boring?

That will depend on your conditioning. If you are excited at the idea of finding out how our Universe works, then you will not be bored in mathematical physics.

If anything, mathematical physics is getting more and more interesting every day, because there are some deep issues in the discipline that no one has much of a clue about. The data from experimental physics and cosmology is piling up in huge amounts with ever better observations on many fronts, yet theory is currently stymied on many of them. I'll discuss a few of them in various places in this book. You may have heard of *black holes*, *String Theory*, *Many Worlds*, quantum mechanics, *dark energy* and *dark matter*. These are all enigmatic concepts in mathematical physics that currently cause mathematical physicists great headaches trying to string together into one coherent, unified theory, the so called *Theory of Everything*. No one knows if it's even sensible to try to do that.

If you're a mathematical physicist, you will have a huge range of intellectual challenges that will keep you occupied for years. One thing is for sure: *you will not be bored.*

## 4.4 Mathematical physics is good for retirement

There's an important advantage in mathematical physics as a career: when you retire, you can still keep doing it, if you want to.

After many decades of employment as a mathematical physicist in one form or another, I found that mathematical physics had done more for me than provide me with an income. It had became a way of life. In consequence, when I retired formally from the maths department where I worked for several decades, I instituted in my home office a regime not much different from the one I had just retired from. I now treat every day from Monday to Friday as a *normal working day.* I try to keep "office hours" and schedule my mathematical physics book writing and research activities over the day as if I was still in post in the maths department. The fact that I don't get paid in actual money is irrelevant. I get paid in many much more significant ways. Three of these are the following.

### *Mathematical physics is good for the brain*

Significantly, since my retirement, I've not relaxed mentally. If anything, I'm busier than ever before. That's really important to me as an individual. I don't want to slow down mentally. I'm writing books and research papers based in mathematical physics. How successful I am at it is irrelevant. Success cannot, after all, be demanded or imposed. It either comes or does not come to us according to factors we rarely have any control over. One of those uncontrollable factors is just being at the right place at the right time. Another is having the right parents or meeting the right inspirational characters.

My point is that work, employment, jobs: these should not always be thought of as bad things designed to make slaves of us, working for rich capitalists or cruel regimes. Viewed from certain perspectives, most jobs can be viewed as opportunities to develop personally.

### *Mathematical physics is enjoyable*

Another way in which mathematical physics has repaid me is in sheer *entertainment value.* This may sound strange. How does doing mathematical physics compare with going to the theatre or watching a film?

Here's where you as an outsider (I assume you're not a mathematical physicist yet) may not appreciate the point I'm going to make. The plain fact is, when I look at some theories devised by mathematical physicists such as Newton, Maxwell, Dirac, Einstein, Schrödinger, Feynman, and so on, I get a feeling of satisfaction, elation, joy, and wonder, an uplift in emotion that beats watching any film or play. That's because those theories are about the real world I live in, about real phenomena I can experience, such as gravity, inertia, temperature, and friction, not fictions of some screen writer's imagination. I can't think of anything more stimulating than that.

I suppose that aspect of mathematical physics has been conditioned into me so deeply that I don't now think of mathematical physics as something I do. It's part of who I am as a person. *I'm not working at mathematical physics, I am a mathematical physicist.* That's been so ingrained into my thinking that I no longer see it as conditioning. As *Ty Webb* says in the film *Caddyshack*:

"*I'm going to give you a little advice. There's a force in the universe that makes things happen. And all you have to do is get in touch with it, stop thinking, let things happen, and* ***be the ball****.*"

### *Something to do naturally, economically, and anywhere*

Another invaluable way in which mathematical physics has served me well is that I didn't have to think of new things to do once I retired. I just kept on doing most of what I had been doing for many decades, apart from travelling to work and back, meeting students, lecturing, and attending meetings. I haven't had to buy any expensive equipment such as a racing bicycle or pay for lessons in a foreign language. There's no end of fascinating problems to work on in mathematical physics in all sorts of areas. I sympathize with people who retire and then find themselves with nothing to do all day. Mathematical physics does not require any great resources. A pad of paper, a good pencil, some peace and quiet, and your brain. That's all you need.

# Chapter 5

# What We Do

*Science, at bottom, is really anti-intellectual. It always distrusts pure reason, and demands the production of objective fact.*

H. L. Mencken [Mencken (1956)]

It's all very well me suggesting that you think seriously about mathematical physics as a career, but I imagine you don't have much of an idea what being a mathematical physicist means. In other words, what is it exactly that mathematical physicists do all day?

That is a simple question with a rather complicated answer. I'll *modularize* my answer, meaning I'll answer in a series of steps or stages.

## 5.1 Where can you find mathematical physicists?

If you walked into a bank, you would expect to find bankers. That's where they hang out. Likewise, if you walked into a police station, you would expect to find police officers[1]. So where do mathematical physicists live?

Mathematical physicists could be found almost anywhere. A couple of hundred years ago, there was a wonderful mathematical physicist called George Green who worked in a wind-powered flour mill. He's one of the heroes of mathematical physics. As you may know, the protagonist of this book, Albert Einstein, is famed for working in a Patent office whilst he was formulating his ideas about Special Relativity. Another great example is Karl Schwarzschild who was serving as a soldier in the German army on the Russian Front in World War One when he worked out the geometry of *black holes*.

More commonly, you will find mathematical physicists in departments of mathematics, where they may actually admit to being mathematical physicists. If they were attached to a physics department, they might call themselves theoretical physicists. If they were working in some other branch of science, such as cosmology or astronomy, they might have an office in an observatory or university department of Astrophysics. It's important to appreciate that a mathematical physicist is a versatile person and, in principle, should be able to fit into any scientific environment where some theorizing is called for.

## 5.2 Solving problems

I'd say that mathematical physicists are primarily *problem solvers* of a certain kind. They don't do Sudoku (unless they're relaxing) or find out who stole the jam tarts. The problems they try to solve are often really big scientific questions, such as *why do things fall downwards?* or *why is there a recoil when you fire a gun?* Some of their problems are not so abstract, such as *how round is our planet actually?* There are some deep questions that they don't even try to answer, such as *does God exist?* or *how can I make a great deal of money?*

As a rule, mathematical physicists are interested in understanding the empirically validated rules by which this Universe of ours actually works.

## 5.3 Making predictions

No, I don't mean letting you know when you will meet a tall, dark, handsome stranger. I don't mean what the weather will be in a week's time either. I mean something far deeper. Let me explain.

This world of ours is immensely complicated. It's made up of a truly vast number of atoms and molecules, which themselves are made up of strange objects such as electrons, protons, and other stuff. Trying to understand how it all hangs together has proven to be quite a challenge over the last two or three **thousand** years or so. So far, knowing something about the laws of physics has turned out to have a lot of benefits to humanity (as well as one or two dangers, such as pollution and nuclear war). That's why science is done.

So far, scientists have managed to work out a lot about those laws of physics. Mathematical physicists are at the cutting edge of research into those laws. Doing that research is as exciting as it can get, way more exciting than playing computer games, sport, and most other displacement activities, but only if you have been conditioned to think about it in the right way.

At this time, some really big unresolved questions remain, and that's where the creativity of mathematical physicists comes in play. For example, James Clerk Maxwell played around (on paper) with the equations of electromagnetism and came up with the theoretical conclusion that electromagnetic effects should travel across space in the form of waves with a specific speed that depended on certain known physical constants. When he put the numbers in to work out what that speed was, it turned out to be *the actual, measured speed of light*, to within experimental error. It's so unlikely to be a coincidence that Maxwell was led to the conclusion/prediction that light *is* no more than an electromagnetic wave effect. That's one of the triumphs of mathematical physicist that stands out as a work of genius.

Of course, there's a bit more to light than just being a wave process. A lot more, in fact. One day, if you become a mathematical physicist, you may well think of light in terms of relativistic quantum field theory. The prediction that light is an electromagnetic effect is an example of what I mean by "making predictions".

Another amazing prediction was by Einstein, who resolved a long standing problem involving the precession[2] of the orbit of the planet Mercury around the Sun. It was known for many years before Einstein's time that Newton's laws of motion and gravitation could not quite explain the observed behaviour of the planet Mercury. When Einstein applied his theory of General Relativity to the problem, he found that there should be just the right, tiny, effect involving the curvature of time and space that would account for the disagreement between Newtonian theory and the observations.

Obviously, in the case of Mercury's orbit, Einstein's calculation was not a *prediction* but an explanation, simply because he explained a known problem. But the wonderful thing is that, given the now overwhelming evidence for the existence of exoplanets (planets orbiting other stars), Einstein's theory can be used to *predict* that the orbit of each one of those exoplanets should also precess around their parent stars, just like Mercury. Moreover, Einstein's theory will predict the precise amount of precession for each exoplanet, as that depends on the masses and distances involved.

A more recent, spectacular example of making predictions is the *Higgs boson*. It was almost fifty years before experimentalists found empirical evidence for it.

## Career Choice Gamble

The experimental detection of the *Higgs boson* in 2012 was a crucial event for mathematical physicists. Many of them had worked for decades on the theory associated with the Higgs boson. If no evidence for it had been found, then it would not be an overstatement to say that it would have represented more than just a theoretical catastrophe for those theorists. Essentially, non-detection of the *Higgs boson* might have led to the conclusion that those theorists had wasted their careers. Like a theoretical *Sword of Damocles*, that possibility still hangs over mathematical physicists who have worked for years on *supersymmetry*. To date (2019), there has been no direct empirical evidence for supersymmetry.

Suppose you started doing a doctorate in a particular field of research when you were say twenty four, at a time when it was a field bursting with enthusiastic researchers worldwide. Suppose further that by your mid-thirties, the field was getting recognized as an empirical dead-end. You would be up the proverbial creek. You could stick it out, working away for another twenty years on that dead-end, or you could jump ship and go into a new field of research. But starting at the bottom in a new field is never going to be easy. That's why I would advise anyone going into mathematical physics to be careful and thoughtful about which field of the subject to go into. Be particularly wary of topics, such as *String Theory* and *Many Worlds*, that are surrounded by a great deal of hype but currently have no critical empirical support.

There's one area of research in mathematical physics that I would certainly jump into if I were now a beginner trying to find a good area of fundamental research:

## The Great Pond

There's an enormous problem currently in mathematical physics concerning the so-called *dark matter*. It was hinted at in the 1930's by the work of the great scientist Fritz *Zwicky*. He is one of the best examples of freethinking you could ask for. He was years ahead of his time, was enormously versatile in his scientific interests, and argued strongly with other scientists about fundamental concepts such as the Big Bang. Fortunately, he did have the position and authority to publish scientific papers that pointed out the existence of a vast unseen distribution of mass of unknown origin in the Universe.

Eventually, other astrophysicists, such as Vera Rubin working on galactic rotation [Rubin (1989)], found more evidence for anomalous and inexplicable gravitational forces that were consistent with Zwicky's *dark matter* conjecture. Astronomical observations are now beginning to come down firmly in favour of Zwicky's disturbing conclusion, that out there in deep space, permeating galaxies and intergalactic space is an enormous quantity of mass that cannot be seen by ordinary (optical) telescopes.

No one knows for sure what *dark matter* is. It could be a type of particle that simply does not interact with light, so it cannot be seen by optical means. That's why it's called *dark*. This is rather like the Earth's atmosphere, which we humans cannot normally "see", except when it's foggy. But the *wind* can be felt, and that's how we believe in "air" as something invisible yet nevertheless quite material. An absence of air, for example, such as under water or in outer space, can certainly be a deadly affair. Like air, *dark matter* makes its presence felt by non-visual means. Specifically, *dark matter* was discovered by Zwicky because collections (clusters) of galaxies (which are vast collections of stars and gas) were behaving contrary to standard expectations, as if there were vast quantities of unseen matter pulling those galaxies together.

Speculations abound at this time. Not everyone thinks *dark matter* consists of "particles" in the usual sense. Dark matter could be some strange behaviour of empty space itself. No one knows right now. There is apparently so much of this material in the Universe that ordinary matter (that is, the stuff that makes you, me, and everything that we can see) amounts to only about 5% of all mass in the Universe.

*That* is clearly an extremely interesting and somewhat disturbing concept. It's as if we were pond insects skating over the surface of a deep deep pond. The optical Universe, the one we could see with our eyes and ordinary telescopes, is like the surface of that pond. Invisible to us, and right under us, would be a much bigger phenomenon, the bulk of that pond.

It's disturbing, because it means we really don't have such a good grasp on what physical reality is. But I suppose if you consider the strange rules of quantum mechanics, you would have probably come to that conclusion without *dark matter* chipping into the discussion.

If I were going into mathematical physics right now, I would choose to research into *dark matter* and dark energy. My reason is that, unlike *String Theory*, *Quantum Gravity*, and *Many Worlds*, there is abundant real evidence for something strange lurking throughout the Universe, something we cannot yet quite "understand" with our current theoretical models. It's precisely under that sort of condition that great mathematical physics can be done, perhaps sweeping away our current scientific conditioned ideas and introducing really new and wonderful concepts. That is precisely what happened when quantum mechanics was discovered.

Understanding dark matter is perhaps *the* great challenge of our age. Perhaps *you* will be the one to come up with a revolutionary theory about it all.

## 5.4 Sub-species

You may well ask: *what's the difference between an applied mathematician, a theoretical physicist, and a mathematical physicist?*

I'll refer to them as mathematical physicists of the *first kind*, *second kind*, and *third kind*, respectively.

Some people might say there was no difference. Others might say that this was a good call for a joke, such as:

**Scottish Cows**

An applied mathematician, a theoretical physicist, and a mathematical physicist are on a train going into Scotland and the first thing they see is a single black cow in a field.

The applied mathematician says "*All cows in Scotland are black.*"

The theoretical physicist shakes his head and says "*No. On the available evidence, you could only say that in Scotland, there is at least one field in which there is a single black cow.*"

The mathematical physicist shakes her head and says "*No. All you can say is that in Scotland, there is at least one field in which there is a single cow, at least one side of which is black.*"

Having worked as a mathematical physicist of the second kind in a physics laboratory (University of Kent at Canterbury) and in a physical chemistry laboratory (University of Oxford), and as a mathematical physicist of the third kind in a mathematics department (University of Nottingham), my experiences lead me to the following (personal) view of the differences between these sub-species of the species *Mathematica Physicus* in the genus *Physicus* (Scientist).

### *Applied mathematicians*

An *applied mathematician* develops and applies mathematical techniques to specific physical problems, such as fluid flow in pipes. Their work often involves making mathematical models, such as representing the earth as an approximate sphere, or a vast collection of molecules as a continuous fluid. Applied mathematicians will probably work in terms of *good enough approximations* to the assumed "true" solutions to problems, where "good enough" depends on requirements. Such theorists are often funded by business and industry, as their work has economic value.

I would also label as applied mathematicians those mathematical physicists who develop and publish purely mathematical technologies unrelated to specific physics problems but are intended to be of value in as yet unforeseen physical situations. For example, applied mathematicians may develop solutions to unusual differential equations, such as the Schrödinger wave equation for a harmonic oscillator in fractional spatial dimensions.

### *Theoretical physicists*

A *theoretical physicist*, as I would use the term, would be a mathematical physicist applying mathematics to a specific experiment, such as predicting the signal response in a magnetic imaging machine. Such theorists will be found in science departments that have attracted funding in cutting edge empirical research.

Here's a joke about how versatile theoretical physicists can be.

**The Graph**

The scene is a coffee room in a physics department. *Theorist* is sitting there drinking coffee when *Experimentalist* walks in and sits down with a sad expression.

"What's up?" asks *Theorist.*

"I've just plotted my data on this graph but I can't understand why the signal goes like this as the temperature increases. Look!" replies *Experimentalist*, pointing to a sheet of paper in his hand.

"Give me a week to explain it," says *Theorist*, snatching the sheet and walking off.

A week later, *Theorist* comes into the coffee room in triumph. "I've done it!" she proclaims. "I can explain the data! Look, my theory predicts that the data should rise as temperature increases!"

*Experimentalist* frowns. "You got the graph upside down! My data *falls off* as temperature rises."

"Give me another week to explain that." says *Theorist*, walking off confidently.

### *Mathematical physicists*

The term *mathematical physicist* could I suppose refer to both of the two types of theorist I've just mentioned, but in my mind, the expression "mathematical physicist" (of the third kind) is more than that. It conjures up an image of a person trying to discover and extend the laws of physics, rather than just applying the known laws to any particular situation. Ideally, a mathematical physicist of the third kind is, to paraphrase the opening lines of *Star Trek (the Original Series)*, "*boldly going where no mathematical physicist has gone before.*"

A good example of what I mean was Schrödinger, whose wave equation shook the foundations of classical mechanics and inspired mathematical physicists to rewrite the laws of physics.

## 5.5 How society views us

Being primarily interested in profit, industrialists will not, in general, fund mathematical physicists of the third kind. Surprisingly, funding for mathematical physics of the third kind comes from the same source as the funding of pure mathematicians, philosophers, and other freeloaders that appear to have no obvious immediate economic benefit to anyone apart from themselves and university finance officers. That source is *Society.* Even in ancient times, societies supported (or tolerated) philosophers and suchlike, because there must be, I think, an unstated collective understanding that societies need them in the long term. Without thinkers, you don't get improvements in technology, for one thing.

For my money, mathematical physicists of the third kind represent the ultimate in that respect. I can give three critically important examples of their work: *classical mechanics*, *electromagnetic theory*, and *quantum mechanics*. These are the foundations of our modern technological civilization and all were developed by mathematical physicists of the third kind **not** under contract to industrialists or business interests.

## 5.6 How mathematical physicists actually work

Great mathematical physics theories such as Einstein's General Relativity, Maxwell's electrodynamics, and Dirac's electron wave equation are great for several reasons. First, they match empirical data excellently. Second, they provide foundations for more theories to be built over them. But perhaps most relevant to us in this chapter, they leave us with a sense of awe, wondering *how on earth did that person ever think of all that?*

Great theories do not appear just like that. They are the products of intense work over extended periods of time. They are the stringing together of vast amounts of knowledge about empirical data and other theories, by stubbornly obsessive individuals who are determined to make sense of it all. Those individuals, such as Maxwell, Einstein, and Dirac, were invariably surrounded by many other, often technically better, mathematical physicists, all of whom bounced their ideas off each other. They would have been aware of their competitors getting close to some answers, and that would have driven them furiously towards some imagined theory still shrouded in the fog of confusion and the mists of uncertainty. The historical fact is that those geniuses worked out their great equations on their own, with no one else telling them how to do it.

The lone genius stereotype is a partial myth and a partial truth. The process of creativity is arguably the greatest mystery about mathematical physics. If we knew how it was done, we could hope to do it again and again. It's what the title of this book alludes to. This mystery is one of the reasons that attracted me to mathematical physics rather than any other field. The more I read about great mathematical physicists, the more I became intrigued about how they came up with their theories.

There is a serious issue in education related to this and that is the group-project/individual-projectee dichotomy. Over the years, I have seen education policies in schools and Universities swing towards the notion that group work is best. I was never convinced of that as far as mathematical physics is concerned, because the historical evidence is dead against the group work paradigm. In the face of well-intentioned but misguided educationalist policies, group work became normalized.

A significant advantage to group work is that it's more economical than having to supervise students individually in their projects. Eight members of staff can supervise eight groups containing five students each in a smaller total number of supervision sessions than are required to supervise forty individual projectees separately.

There are two points that counter the group-project mentality. First, throughout my decades of life in Universities, there have always been departments and sub-departments that specialized in mathematical physics. Their student intakes have always been relatively small, compared to subjects such as physics, chemistry, engineering, and, surprisingly, mathematics itself (which continues to attract large numbers of students). It's as if, despite the fluctuating attitudes in education policy, society has blindly recognized the need for the specialist subject of mathematical physics.

The second point, for which I have found some evidence for, is that as far as creativity is concerned, *group projects don't seem to work.* You can find numerous articles online that discuss this issue. Some of them conclude that there are serious questions about the value of group work regarding creativity [Walton (2016)].

My view is that students should do *some* group project work, but they *must* also have a go at individual projects involving only themselves and a supervisor. This latter scenario is precisely what happens during a doctorate, which, as far as I know, never involves a dissertation written by a group.

Mathematical physics, at its best, involves creativity. Group interaction inevitably brings in common sense points of view and a cautious attitude to taking risks. Creativity should not be expected in those circumstances. My best advice to the budding mathematical physicist is to read widely, talk to people as much as possible, but go find your own corner somewhere where they can't find you, and *work out your ideas on your own.*

## 5.7 I've got a great idea ...

This is perhaps a strange point to put across, but it has to be done. *Don't expect anyone to do your work for you.* The following anecdote tells it all.

**Death Ray**

Some years ago, I came across a Sunday newspaper supplementary magazine with the following story.

During the Second World War, times were hard. Battles raged, people died, and the outcome of the war was uncertain.

Up and down the land, great efforts were made by civilians to assist the military as much as possible. This occurred of course in every country involved. Many people grew vegetables in order to assist food production. Some people helped on farms. Others went around collecting metal, to help build guns and tanks. In the village where I live, there is a house with a front garden. There is no fence. All you can see are the metal pieces at ground level, where eighty or more years ago, someone cut the metal fence off to give towards the production of weapons. That was a common sight.

Some people tried to help by giving advice. The newspaper magazine article I read was about a letter sent to the Ministry of War (in those days), giving detailed plans for a *death ray* weapon that would certainly give victory.

The letter started by saying that the weapon would be carried in an airship towards its target city, and there it would be activated and the city would be destroyed. The author of the letter gave a fully comprehensive list of requirements:

(1) One standard airship with crew of six.

(2) 300 gallons of petrol for the round trip, there and back, to the enemy capital.

(3) Food and water for six crew to last the three day journey.

(4) Sufficient air cover to ensure adequate protection of said airship and crew during mission.

(5) *One death ray, which the Ministry surely already has.*

On a number of occasions, I was approached by some person who would open the discussion by suggesting that, as a theorist, I wasn't doing anything particularly urgent or important. On the other hand, *they* had some really important ideas, blah blah blah. *All* that was needed was for someone (meaning *me*) to work out the theoretical details, which shouldn't take long. Wouldn't I like to drop whatever I had been working on these past five years and immediately work out their theory for them?

I trust you see what I'm driving at. There's a vast gulf between having what you think is a great idea, and actually encoding it properly into a sound mathematical theory. Do not expect anyone to do that encoding for you, unless you either pay them or have the personality to smooth your way into their affections. Theorists are always busy with their own ideas. They don't need anyone to saddle them with half-baked proposals for time-machines, death rays, and such like.

The moral is: *if you have a great idea, prove it's great by* ***doing it yourself.*** That's what Maxwell, Einstein, and Dirac did.

## Chapter 6

# Nullius in Verba

*Mind tricks don't work on me*

Watto, Star Wars Episode 1 (*The Phantom Menace*)

The title of this chapter is a Latin saying. Latin is a dead language, so does that mean this saying is a bit out of date?

I don't think so. Those ancient Romans were pretty smart people. They experienced many of the issues that we have today. They found that sayings, mottoes, and proverbs were a good way of reminding themselves about those issues, about how they solved them, or ran away from them. We still do that, because we have all of the same issues that they had, plus a few more, I'm sure. Whilst this motto is expressed in a language that dominated Europe for over two thousand years, its significance to us dates from about 1660, when the *Royal Society* of London was founded.

*Nullius in Verba* can be an extraordinarily useful motto, if you understand what it really means and apply it to your scientific work and to other aspects of your life. I came to realize the immense power of *Nullius in Verba* only a few years ago, when I started to write my books. It is applicable not only to scientific research but to many aspects of human life. I think it is one of the most important sayings anyone could use on a daily basis. I use it all the time now and it stops me from making some big mistakes, believe me.

If you do come to understand its real meaning, you'll realize that, whilst it has immense benefits, it could be dangerous to use openly in some circumstances, because it challenges baseless authority.

## 6.1 Translation of *Nullius in Verba*

The exact translation of *Nullius in Verba* has been discussed at length [Dawes (2012)]. Some care is needed, because a literal translation does not convey the historical intent of the motto. For instance, translated into English via *Google Translate*, it reads "*no words*". That's a poor translation, because it doesn't convey the sentiment (or meaning) behind the actual Latin. *Nullius in Verba* is going to be pretty important to you if you become a mathematical physicist, whether you like it or not, so we need to understand more precisely what it is supposed to mean.

## 6.2 Origins

To understand *Nullius in Verba*, we can do no better than look at its origins. That means understanding the rise of *scientific societies*. These are club-like organizations set up in various countries throughout the world to promote the advance of science and the scientific way of life. Some scientific societies were set up hundreds of years ago and others much more recently. The earliest I can identify as a scientific society consistent with the principle of *Nullius in Verba* was the *Accademia dei Lincei* founded in 1603 in Italy. Galileo was a member of this relatively short-lived society. That society was followed by the formation of a few more private societies, such as the *Accademia del Cimento* in Italy, and then by significant national societies, such as the *Academie des Sciences* in France in 1660 and the *Royal Society* in London just after that.

The Seventeenth Century was by most accounts a brutal time in Europe. The Catholic Church exercised a powerful stranglehold on scientific development throughout most of that time, resisting by forceful means lines of research that appeared to challenge its authority. For example, Giordano Bruno was burnt at the stake in 1600 for his freethinking views about cosmology. Galileo came close to suffering the same fate. The Protestants were not much different: Newton had to hide his views about the nature of God (he was a fellow of Trinity College, Cambridge but didn't believe in the Trinity). The terrible *Thirty Year's War* lasted from 1618 to 1648, demonstrating the fact that conditioning, or what people think and believe, can lead to the most dreadful of conflicts. I'll throw in for good measure the fact that the last (alleged) witch was executed in Britain in 1727, long after

*Nullius in Verba* was adopted in the same country. That is a measure of how difficult it is to change any society's conditioning.

Given all that mayhem and plain stupidity going on, we can only reflect with astonishment that some educated people in those days had the courage to form societies that had the express purpose of conducting proper scientific enquiry into how the Universe worked, free from the imposed authority of religion and superstition. According to the *Royal Society*'s web site,

*Nullius in Verba is an expression of the determination of Fellows to withstand the domination of authority and to verify all statements by an appeal to facts determined by experiment.*

*Nullius in Verba*, therefore, expresses a cardinal scientific tenet: *Don't take anybody's word at face value, whatever they say and whoever they are. Find for yourself real, impartial, objective evidence for anything you want to believe. Don't say things you cannot substantiate empirically.* I could go on, but I think you'll see what *Nullius in Verba* is all about. Once you grasp its meaning, you may suddenly find yourself in a potentially dangerous position. *Nullius in Verba* could be applied not just to scientific research, but to politics, religion, culture, tradition, and everything else that humans do.

It's potentially dangerous, because what happened to Giordano Bruno still goes on around the world: in some countries, little girls still get shot in the head on school buses simply for going to school. My best advice is: be careful. Absorb the lessons that *Nullius in Verba* teaches, but you don't need to shout them from the rooftops. It's safe in many parts of the world at this time, but in some parts of the world, however, advertising the fact that you believe in *Nullius in Verba* could get you into real trouble.

## 6.3 Related mottoes and razors

An online search reveals several sayings that have the same or similar intended point to make. One in particular, usually referred to as *Hitchens's razor* gives you an immediate riposte when confronted with worthless statements: *What can be asserted without evidence can also be dismissed without evidence* [Hitchens (2008)].

A Latin version of Hitchens's razor reads *Quod gratis asseritur, gratis negatur*, which translates as *What is asserted freely can be dismissed freely.*

It's possible to over do reliance on mottoes. I'm thinking here specifically of *Occam's razor*. That's a philosophical principle which says that *if there are two explanations of the same phenomenon, the one with the least number of assumptions is probably the correct one* (or words to that effect).

Occam's razor is, in my view, a suitable candidate for *Nullius in Verba*. I have two concerns with Occam's razor.

First, it's not hard to find issues in science where it is an empirically false principle. An example is the long-standing question of the structure of matter. Historically, there were two incompatible models. The *continuum theory of matter* asserted that matter is continuous on all scales. That's compatible with the false *Fourier principle of similitude*, which asserts that the laws of physics are the same on all scales. The other model, the *atomic hypothesis*, asserted that at a certain scale, matter is composed of discrete units known as atoms.

My point is that a pre-1900 philosopher looking at this question would assert that a piece of wood is divisible into halves, then quarters, then eighths, and so on, *ad infinitum*, the only problem being in finding knives sharp enough to keep dividing the wood for as long as we wished. The philosopher would assert that there was no natural scale on which the wood no longer looked like a continuum but was composed of atoms. The simplest assertion, therefore, would be to assert that wood is continuous.

Of course, the pre-1900 philosopher would know nothing about Planck's constant, the quantization of electric charge, and definite electron mass, all of which contribute to the scale of atoms, invalidating the continuum argument.

The continuum-atom debate was not a pleasant one. The mathematical physicist Boltzmann developed *statistical mechanics*, which is based on the atomic hypothesis, and accounted for the classical laws of thermodynamics in gases. He faced extraordinary heavy criticism from philosophers and some scientists, such as Ernst Mach, who did not believe in atoms. Boltzmann tried unsuccessfully to argue against the assertions of the continuum theory proponents until eventually, he became depressed and killed himself. I cannot help the thought that a rigid application of *Nullius in Verba* would have helped him in the debate.

The second problem I have with Occam's razor is that it suggests that there is some sort of 'truth' out there, and that it will in general take the form of simplicity. I don't think there is an absolute truth there at all. In physics, there are only empirical, contextual truths. If they happen to look ugly, so what, if that's what *Nullius in Verba* leads to.

Moreover, today's mathematical physicists have a rich, complex, and successful theoretical understanding of a lot of the Universe. But, it's not complete, and it's not particularly elegant or simple in places. I'm thinking of the *Standard Model* here. It works, more or less, but it's not perfect. The dangerous complacency of Occam's razor has leached into certain branches of mathematical physics, however, such as *String Theory* and *Many Worlds* leading numerous mathematical physicists to believe that nature has to be beautiful and simple at heart. That's just an assertion that has looked true at certain times, but need not in principle be true in a fundamental way.

I think any budding mathematical physicist would do well to seek elegance and mathematical beauty in their equations, if possible, but to be wary of assuming that theories have to be like that. In other words, I would not trust Occam's razor as anything more than a guide.

## 6.4 Relevance to mathematical physics

Given a belief in the principles of *Nullius in Verba*, how should you apply them to mathematical physics?

My view is this. Never forget that the term *mathematical physics* contains the word *physics*. It seems to me that some mathematical physicists have forgotten or ignored that basic fact. If you're not careful, you may find yourself doing research into topics such as *String Theory* and *Many Worlds*, which currently (2019) have no critical scientific evidence in their favour.

Why is this important? Suppose such research areas go on for another forty or fifty years each with the same level of investment in time, resources, and careers, with no forthcoming critical scientific evidence. Anyone who had been working in those areas would have been essentially wasting their time. If you go into mathematical physics and want to have a life-long career, you should look carefully at this issue before taking any serious steps. My best advice here is: use *Nullius in Verba* to form your decision in this respect and you will not go far wrong.

## An Old Man's Game

Sure enough, I found myself in the very position I've just discussed. I was approaching the end of my undergraduate mathematical physics degree and wanted to do a doctorate (PhD) in the subject. I had done some background reading and had thought about Relativity and quantum mechanics as possible topics.

I wrote to various departments in the UK and went to a couple of interviews. Such interviews serve two purposes. First, they allow people interviewing you to see what sort of person you are. But equally importantly, they give you a chance to see what sort of quagmire you might be getting sucked into if you went there and spent three or more years doing research on some duff subject.

I looked into Relativity. One particular interview (I won't say where) was really useful. They were developing a theory of spacetime based on *rational* coordinates, rather than just real number coordinates. You will recall that a rational number is expressible as the ratio of two numbers, such as 345 over 8761. A number such as square root of two is not rational, so is excluded in the coordinate system they were proposing to investigate.

Once I understood what the research topic was about, I quickly backed out. It seemed obvious at the time to be a bit of a lemon idea. In particular, I thought that *rotations* as we know them could not be supported in a three-dimensional space where irrational number coordinates were ruled out of order. But then, I suppose research is all about looking at new ideas and finding out how good or bad they are. The problems with the rational coordinate idea seemed to me to be intuitively obvious right from the beginning, rather than after lengthy calculations. After my interview, which I think went well, I gave that line of research a miss.

I then thought about quantum mechanics. I found out that there was a department doing research into Hidden Variables. That's the idea that all the weirdness and non-classicality of quantum mechanics could be circumvented by assuming that underneath the superficial layer of empirically observable reality seen by our detectors in the laboratory (the layer that needs quantum mechanics to describe it), the real world is actually inhabited by unobservable variables that are classical all along.

My view of this now is that it's a good case for treatment by *Nullius in Verba*. You say there are unobservable hidden variables? You say that, by definition, no one can actually detect them? You have no direct evidence for that belief, but nevertheless, you think you can "explain" everything.

Pull my other leg as well. Over the years, I've thought all about this matter a lot. What difference is there between believing in *Hidden Variables* and believing in Divine Providence, or Fairies, or Magic, or whatever takes your fancy? Just because you say *Hidden Variables* is scientific does not make it scientific.

I have to be careful here. People considered the atomic theory of matter to be nonsense but it turned out to be correct. If hidden variables exist, then they should be detectable eventually. Otherwise, you're just playing with metaphysical ideas that have no place in science. I would have no quibble with the *Hidden Variables* idea, if people would actually propose *realistic, doable* experiments to detect them. But unfortunately, the real point about hidden variables is that they cannot be detected. So the idea will always be metaphysical and not scientific.

The greatest problem you face when you're young is that you just don't have the sophistication that experience gives you, because you haven't had much experience. It seems reasonable when you're young to believe in a hidden classical reality underneath all the jibber-jabber of quantum mechanics. So I started to think about applying to have an interview at that department doing Hidden Variables.

Fortunately, I decided to ask some lecturers what they thought about *Hidden Variables*. I went to see Peter Higgs. This was many years before he got his Nobel Prize. What he said put me straight, and it's a good piece of advice I want to pass on to you now:

**The interpretation of quantum mechanics is an old man's game**[1].

I decided not to go to that interview, and for that, I will always be grateful.

My interpretation of that advice is this: *Some ambitions are impossible. "Understanding" quantum mechanics is probably one of them. Don't waste your time trying to understand quantum mechanics when you're young. When you're a clapped out, wizened old mathematical physicist, then, by all means, potter about having a go tilting at that particular windmill.*

After many years, my view of quantum mechanics is that it's *us* that's wrong. It's our classical conditioning that gives us expectations that are not fulfilled precisely by what we see in the laboratory. Understanding how humans think is, I would say, one of the great remaining challenges in science, along with understanding dark matter and life itself.

## 6.5 Is *Nullius in Verba* a recipe for anarchy?

At first sight, *Nullius in Verba* sounds like *anarchy*, a challenge to order and authority. If you understand the true meaning of *Nullius in Verba* however, you'll know that to be a mistaken view. Anarchy is a state of lawlessness. In an anarchic society, individuals are free to do whatever they want, even if it means harm to others. In contrast, *Nullius in Verba* is a recipe for deciding which laws to obey. *Nullius in Verba* tells us not to obey the baseless whims of dictators, religious fanatics, and pseudo-scientific charlatans, but to accept laws that are based on objective evidence.

I imagine the greatest objections to *Nullius in Verba* to come from those who would benefit most by its neglect.

## 6.6 The use of mathematics

Whilst reviewing this chapter, it occurred to me that I should mention the role of mathematics in mathematical physics. There is a paradox in mathematical physics. *Nullius in Verba* tells us not to believe in statements that cannot be verified empirically. But mathematical physicists have no choice but to use *mathematics*, which is entirely to do with concepts such as numbers, wave functions, quantum fields, gauge transformations, Hilbert spaces, and a plethora of other ideas, all of which exist only in the human imagination. Where does imagination end and physical reality begin? Why is mathematics apparently necessary in mathematical physics? And why is it so good? No one really knows the answers here. Perhaps *you* will have some insights into all this.

# Chapter 7

# The Gates of Hell

*We knew the world would not be the same. A few people laughed, a few people cried, most people were silent. I remembered the line from the Hindu scripture, the Bhagavad-Gita. Vishnu is trying to persuade the Prince that he should do his duty and to impress him takes on his multi-armed form and says, "Now I am become Death, the destroyer of worlds." I suppose we all thought that one way or another.*

Robert Oppenheimer [Temperton (2017)]

## 7.1 Didn't mathematical physicists invent atom bombs?

I start this chapter with a quote attributed to the mathematical physicist Robert Oppenheimer, who witnessed the fireball rising from the first (test) detonation of a nuclear bomb, the so-called *Trinity nuclear test* in 1945. I never planned on writing such a chapter as this one, but because I want to encourage students to go into mathematical physics, I cannot avoid discussing some dark issues to do with the discipline.

Mathematical physicists *were* directly involved in the creation of nuclear bombs in three crucial ways. First, they were the ones who stumbled on the laws of physics that could be used to unlock tremendous energy from ordinary matter. Einstein had a leading role in this respect, because he discovered the rule $E = mc^2$ for converting mass into energy. Second, they realized that such energy could be organized and concentrated in such a way as to create tremendous explosions. Einstein wrote a famous letter to the American President in 1939 saying precisely that. Finally, mathematical physicists helped design bombs that actually worked and destroyed the cities of *Hiroshima* and *Nagasaki* at the end of the Second World War. As far as I know, Einstein had no part in this third phase.

It's an undeniable fact that if mathematical physicists had not discovered the laws of sub-atomic physics, no one would ever have built an atom bomb. My defence is that it's *politicians* who decided on the construction of those bombs and on their use. The mathematical physicists who pointed out the possibility of making those bombs were frightened that the Nazis would make and use them first.

No one is spared when it comes to issues of morality. Einstein sent the following letter to the American President in 1939, just before the Second World War broke out, warning him that an enemy country might develop atom bombs and suggesting that America start building such bombs itself:

**The Letter**

Albert Einstein
Old Grove Rd. Nassau Point
Peconic, Long Island
August 2nd, 1939

F.D. Roosevelt
President of the United States
White House
Washington, D.C.

Sir:

Some recent work by E. Fermi and L. Szilard, which has been communicated to me in manuscript, leads me to expect that the element uranium may be turned into a new and important source of energy in the immediate future. Certain aspects of the situation which has arisen seem to call for watchfulness and if necessary, quick action on the part of the Administration. I believe therefore that it is my duty to bring to your attention the following facts and recommendations.

In the course of the last four months it has been made probable through the work of Joliot in France as well as Fermi and Szilard in America-that it may be possible to set up a nuclear chain reaction in a large mass of uranium, by which vast amounts of power and large quantities of new radium-like elements would be generated. Now it appears almost certain that this could be achieved in the immediate future.

This new phenomenon would also lead to the construction of bombs, and it is conceivable-though much less certain-that extremely powerful bombs of this type may thus be constructed. A single bomb of this type, carried by boat and exploded in a port, might very well destroy the whole port together with some of the surrounding territory. However, such bombs might very well prove too heavy for transportation by air.

The United States has only very poor ores of uranium in moderate quantities. There is some good ore in Canada and former Czechoslovakia, while the most important source of uranium is in the Belgian Congo.

In view of this situation you may think it desirable to have some permanent contact maintained between the Administration and the group of physicists working on chain reactions in America. One possible way of achieving this might be for you to entrust the task with a person who has your confidence and who could perhaps serve in an unofficial capacity. His task might comprise the following:

a) to approach Government Departments, keep them informed of the further development, and put forward recommendations for Government action, giving particular attention to the problem of securing a supply of uranium ore for the United States.

b) to speed up the experimental work, which is at present being carried on within the limits of the budgets of University laboratories, by providing funds, if such funds be required, through his contacts with private persons who are willing to make contributions for this cause, and perhaps also by obtaining co-operation of industrial laboratories which have necessary equipment.

I understand that Germany has actually stopped the sale of uranium from the Czechoslovakian mines which she has taken over. That she should have taken such early action might `perhaps be understood on the ground that the son of the German Under-Secretary of State, von Weizsacker, is attached to the Kaiser-Wilhelm Institute in Berlin, where some of the American work on uranium is now being repeated.`

`Yours very truly,`

`Albert Einstein`

It's a historical fact, then, that many mathematical physicists, including Einstein, *were* directly or indirectly involved in the development of atom bombs during the Second World War and for years after that. Some of them are still doing it.

## 7.2 Justification?

Can such work ever be justified?

Standing now safe and sound, many years after that conflict, it's easy to say that such work is a perversion of what mathematical physics is supposed to be all about. By all accounts, Einstein was a pacifist, yet he wrote that letter. That illustrates the dilemma of pacifism: how far do you go to prevent harm to others? I'm fortunate I was never put in the position of many mathematical physicists during the Second World War that were faced with that question. I did not intend this book as a guide in morality. If you become a theorist, you may well find yourself face to face with such issues as these. Make sure you know which sort of theorist you want to become. You will want to sleep at night in your old age.

The following anecdote is about a story I heard many years ago.

### The Quark Bomb

In the early years of the twentieth century, mathematical physicists realized that in order to get some sort of understanding of the structure of matter, it was necessary to smash it to bits first, and then look at those bits separately. That's where the concept of *atom smashers* comes from.

Another name for an atom smasher is *particle accelerator*. This a machine that throws objects together at great energy and then allows for an analysis of what happens after that. Particle accelerators were used throughout the Twentieth Century to gain great insights into the fundamental laws of physics.

It wasn't that long ago, in historical terms, that most people didn't believe in atoms. After all, you can go pretty far in cutting a piece of wood into two halves, then cutting one of those halves into two halves (meaning two quarters of the original), and so on forever.

Forever? Really? How do you know? Just how long *could* you go on cutting those ever smaller pieces of wood?

What happened over the years was that as particles were thrown together at greater and greater energies, more and more strange properties of matter were revealed. But this process is difficult for several reasons. *Energy* is one of them. As physicists investigated deeper and deeper into the structure of atoms, their particle accelerators had to accelerate particles to higher and higher energies. That leads to perhaps the biggest problem of all: *money*. You need lots of it to do modern particle physics.

In my first year of PhD, I attended a lecture from a visiting theorist on the work being done at *SLAC*, the *Stanford Linear Accelerator* as it was known in those days. I took copious notes. Never having been there, I could not relate to what the speaker said. Has such a thing ever happened to you?

I wrote down what I heard. When the speaker said "*SLAC*", I headed my notes on the talk *SLACK* (with a 'k'). In my mind, I pictured a load of physicists loafing around, drinking beer. Imagine how I felt when a fellow research student noticed and put me wise.

I found out more about such particle accelerators, about how great new physics would come from them. That certainly was correct in the long run. But the real problem, as with just about everything, was *funding*.

It was then I heard one of my fellow graduate student relate the apocryphal story about particle physicists in America who had applied to some Government agency to fund a super new high energy particle accelerator. Money was tight and, coming before funding committees, the scientists came under a lot of pressure to justify what appeared to be totally abstract research. Why smash particles together? What good would it do anyone?

The story went that the physicists were getting nowhere with their funding application, until a well-known mathematical physicist, being interviewed by a funding committee and asked to justify the enormous expense, just happened to mention the phrase '*quark bomb*'. And lo and behold, the money was forthcoming.

This anecdote is true in the sense that I did hear that funding committee story. I don't know if that story itself is true. But it doesn't matter. Mathematical physicists proved during the Second World War that what they researched into had the potential to destroy the planet. That, apparently, gives them an importance they never expected.

I've mentioned in another chapter that currently, there is a great mystery surrounding gravitation out there in deep space. There's increasing evidence that there is an unfamiliar form of gravitating matter known as dark matter filling space. Perhaps it has properties that we simply cannot imagine right now. Perhaps *you* will be the mathematical physicist who finds a good theory to understand the stuff. Perhaps that theory would allow the construction of weapons far more terrible than the ones we have now. Then you may need to decide what sort of person you are: someone who opens *Pandora's Box* and lets out all the evils, or else you decide to burn your notes and keep quiet.

The dilemma is: if you choose the latter course of action, you can be sure that sooner or later, someone else will stumble on an equivalent theory and opens that box.

# Chapter 8

# Why?

*An individual has not begun to live until he can rise above the narrow horizons of his particular individualistic concerns to the broader concerns of all humanity. And this is one of the big problems of life, that so many people never quite get to the point of rising above self. And so they end up the tragic victims of self-centeredness. They end up the victims of distorted and disrupted personality.*

Martin Luther King, Jr., [M. L. King (1957)]

One of the things I've noticed over the years about people is how often they do things for no apparently sensible reason.

By *no apparently sensible reason*, I mean doing something that is perhaps hard or even dangerous, for no obvious personal reward, when it might seem more logical not to do it. For example, why do fire officers risk their lives going into a burning building when everyone else is running out? Why does an athlete train for the Olympics without taking drugs, even though they know some of their competitors are almost certainly doing so?

It's not much of an explanation to say that they do it because they "want to". That gives no more understanding as saying that objects fall downwards because of gravity. Yes, but what's gravity?

In this book, I'm trying to motivate students into going into mathematical physics. As soon as I started writing, I realized that I would have some difficulty in getting readers to want to do that. Giving rational, logical reasons for doing hard things is often counter-productive. The successful Roman army was not founded on convincing its legionnaires about the merits of Empire, but on long-term training, harsh discipline, and conditioning to obey orders without question. I realized, as I started writing, that the key element I had to focus on was *conditioning*. I don't think I could justify you going into mathematical physics rather than into much better paid career of investment banking, if you've been irreversibly conditioned to think in terms of financial reward and material benefit.

I don't believe it's a lost cause. If I can get you to recondition yourself, meaning, change the way you think about things, then you may well find yourself passionate about mathematical physics.

Is it possible? Perhaps. It may be an uncomfortable fact to accept, but there's scientific evidence for the statement that just about everything we humans do is influenced either by hard-wired genetically determined programming, such as breathing, eating, and sleeping, or governed by the dictates of conditioning. If we can change that conditioning, our belief structures may change in consequence.

I don't think it's the other way round. I don't believe now (I did once) that our conscious beliefs dictate everything. Here's a rather disturbing anecdote that suggests consciousness is not all that we think it is and highlights the importance of conditioning.

## 8.1 You're not really you

### The Greatest Myth

About twenty years, I bought a copy of *New Scientist*. It's always been a favourite magazine of mine, as it reviews current developments in most branches of science and technology. Sometimes the reporters sensationalize their assigned topics. For instance, they will discuss crackpot interpretations of quantum mechanics as well as review sensible interpretations. Which is nonsense and which is sense is down to the reader to decide, ultimately. That's all part and parcel of being properly educated. It's the responsibility of the reader to decide what to rely on or to dismiss as unsound.

In that particular issue I found an article called *The Greatest Myth* by two brain researchers (neuroscientists) called Peter Halligan and David Oakley. You might be able to find this particular article if you can find the relevant back issue of the *New Scientist*. To help you find it, here is the reference [Halligan and Oakley (2000)]. As a mathematical physicist you will find yourself tracking down references a lot.

What Halligan and Oakley say in their article is that we humans are not what we think we are. When we make decisions, it seems to us at the time that we have made those decisions rationally, consciously, and in complete control of our decision making processes. Halligan and Oakley discuss the hard scientific evidence that in reality, *decisions are made* ***before*** *we think we have made them.* In other words, our actions are determined by subconscious processes. Our consciousness then steps in and gives us the impression that we are responsible for our actions.

I don't know about you, but that frightened me when I read it all those years ago, and it still frightens me now. It's a scientific vindication of the idea that our conditioning (which operates on our subconscious levels) is perhaps all we've got and are. We are not in control of ourselves as much as we think we are.

By the way, normally I don't value pseudo-scientific discussions about subconsciousness versus consciousness, except when it comes to hard science. Halligan and Oakley were reporting hard science facts that are directly relevant to this book.

What I take from Halligan and Oakley's article is that if we want to achieve something, then we should make sure to deal properly with our conditioning. They do not say in their article that conditioning is permanent. All evidence points to the contrary. You go to school to be conditioned into "knowing" certain subjects such as physics and maths. You go to church, mosque, synagogue, or temple, to be conditioned into definite religious beliefs[1]. You train yourself in a sport so as to play that sport as if it were instinctive. You practice that piano or violin four hours a day and end up playing it from memory and without notes, as if it were second nature to you. I believe it's possible to condition yourself into mathematical physics. But only if you have made a conscious decision to do so. That's what this book may help you to do.

It's possible to overstate conditioning. Humans are not entirely governed by their conditioning. That's really what the debate about free-will is about. If humans had no consciousness, then we as a species would be locked into some unchanging patterns of behaviour, ruled by instinct. Unlike other species, however, we humans have the ability to imagine potential alternative futures, review those possibilities, and make decisions that can alter the way we think.

### Fencing

No, I don't mean fixing boundaries to fields. I mean the sport of fencing. At a late stage in my life, I took up that sport for exercise.

Fencing is a great sport. I like it for a number of reasons. First, fencing allows you to hit another person with a long metal stick, legally. That's a primitive feeling but an immensely satisfying one and it does no harm, usually.

Second, the outfit looks great. I always imagine I'm Zorro or d'Artagnon, fighting the evil Duke.

Third, the action is intense and you get a real workout. I thoroughly recommend it.

The point here is that if you think too much (overthink) what you're doing in a fencing match, you will come a cropper usually. By training, you learn to react without apparent conscious effort. Your fencing coaches will train you to react subconsciously rather than by taking conscious decisions.

Fencing is a sport that caters for just about everyone. You can be of school age or retired. You can be male or female. I imagine you could even do it if you were in a wheelchair.

Fencing is an excellent sport to discuss in the context of Halligan and Oakley's article (discussed above) because moves in fencing are over in fractions of a second, clearly comparable to the time it takes to think consciously.

There is an interesting side to fencing I should mention. As you develop your skills, you progress from a complete beginner, then a novice (partially trained fencer), and finally a veteran (fully trained fencer). I've been told that a veteran should usually beat a novice hands down, but can come a cropper with a complete beginner. That sounds wrong, does it not?

No, it's a logical outcome of over-conditioning. A complete beginner has not had time to accumulate any in-built reactions. They're *fencing unconditioned.* In a match, they will flail their sword arm around almost at random. On the other hand, a veteran fencer will be fully conditioned, to the extent that their subconscious will assume their opponent is going to do something predictable. The veteran's subconscious formulates a standard response that should work against an experienced fencer, but does not allow for the complete beginner's unpredictability. Without knowing what they're doing, a complete beginner could end up beating a veteran fencer. Much the same thing happens in guerrilla warfare, where a well-trained, conventionally based army can be tied down by a much smaller force of insurgents deploying unorthodox tactics.

The question remains: *why should you want to be a mathematical physicist?* I'll give some answers that work for me.

## 8.2 Awareness of conditioning

I've discussed *Nullius in Verba* in Chapter 6. It's a principle that alerts me to the fact that I'm constantly the target of attempts to condition me. *Do this! Buy that! Believe this! Learn this! Don't listen to them!*

*To do as they say, or to think for yourself*? That is the question. I made up my mind a long time ago that my choice was to think for myself. I decided that I wanted to be aware of when I was being manipulated. Only in the last few years have I learnt that *Nullius in Verba* amounts to that decision.

You too have the same choice. If you choose *Nullius in Verba*, then you will be putting your foot on the first step towards a more thoughtful and, in my opinion, better, way of living your life. It may not lead you all the way to becoming a mathematical physicist, but I believe you will see your life in better terms.

## 8.3 It's enjoyable

In my experience, mathematical physics is immensely enjoyable. I have found great satisfaction in working out some complicated problem, writing a paper on it, and having it published in a journal. It's much like uploading a YouTube video and getting lots of hits. Why does anyone do that? Certainly, having a Youtube hit can pay off financially, but there's a bit more to it than that, I'm sure. People who have lots of followers on Youtube must surely feel a great sense of personal achievement. It's natural to know that other people think you're smart or good at what you do.

Of course, as with YouTube, you will get your critics when you publish a paper and someone doesn't like it. When that happens, it can be pretty bruising to your ego, let me tell you. But that's no different (as far as your feelings go) to losing a football match, or coming third in a race. Next time, you may win.

## 8.4 It can be immensely satisfying

There are aspects of mathematical physics that you might not be aware of. If you are so fortunate as to stumble on some great new theory that actually works well in confronting hard data, you may experience something that Richard Feynman felt.

Feynman was an outstandingly creative mathematical physicist who made lasting contributions to quantum mechanics. In particular, I should like to mention his *path integral* approach to quantum mechanics. In that approach, a particle going from $A$ to $B$ is imagined as going in all possible paths connecting $A$ and $B$ *simultaneously*. If that sounds bizarre to you, that's because it *is* bizarre. I can't explain in more detail here and I've simplified the discussion somewhat, but nevertheless, Feynman's path integral approach sums up why quantum mechanics is at odds with our classically conditioned view of the world.

Here in Feynman's own words is what happened one night, after intense thinking about a particular particle reaction known as *beta decay*:

*That night I calculated all kinds of things with this theory. The first thing I calculated was the rate of disintegration of the mu and the neutron. They should be connected together, if this theory was right, by a certain relationship, and it was right to 9 percent. That's pretty close, 9 percent. It should have been more perfect than that, but it was pretty close.*
*I went on and checked some other things, which fit, and new things fit, new things fit, and I was very excited. It was the first time, and only time, in my career that I knew a law of nature that nobody else knew ...*

Richard Feynman [Feynman (1985)]

Just imagine yourself in Feynman's position, having worked out some theory, knowing that right there and then, *you* and only you had that particular insight into the nature of reality. I think every mathematical physicist would love to be in that position. Comparable situations can be imagined as having occurred throughout the history of mathematical physics. Imagine the tremendous elation felt by Maxwell when he calculated the speed of his theorized electromagnetic waves and it turned out to be the speed of light. Imagine the tremendous satisfaction felt by Einstein when he used his theory of General Relativity to calculate the precession of the orbit of the planet Mercury and got an answer that agreed with experiment. Imagine the joy felt by Emmy Noether when she figured out that the conservation of energy and momentum was directly linked to the symmetries of space and time. Imagine the elation felt by Dirac when he finally worked out his great equation for the relativistic, spinning electron.

## 8.5 It's inclusive

An important fact about mathematical physics is its internationalism. There are no barriers. As a mathematical physicist, you will meet talented people from all over our planet who believe in the same principles as you do and have the same ambitions. As a mathematical physicist, you may well find yourself attending a conference in some exotic part of the world that you would never have thought about visiting normally.

## 8.6 It's transportable

An understated side to mathematical physics is its transportability. What I mean is this. You'll often encounter someone playing a guitar or clarinet on a street corner, collecting money from passers-by. They're called *buskers* in Britain. You won't see buskers playing grand pianos on street corners (at least, I've not seen any), because grand pianos are not readily transportable.

The same problem crops up in science. An experimentalist is generally pinned down to their laboratory. If you've ever seen the *Large Hadron Collider* (LHC) at CERN in Switzerland, you'll know that it's measured in kilometres. Scientists have to go to the LHC to work, unless they are theorists. A mathematical physicist needs only a scrap of paper and a pencil to do their work and they can do that anywhere. It's this transportability that led me to create my *mobile office*, which I discuss in Chapter 14, *Tools of the Trade*.

## 8.7 It's for posterity

Here's a rather sad reason for doing mathematical physics. It's about *legacy*, meaning what you leave behind you when you're gone.

It's a fact that most of us will be born, live out our lives in a conventional way, die, and then be forgotten within a couple of generations by our descendants. There's nothing wrong with that. *That is the order of things*, in the words of *Remata'Klan* [Alpha (2374)].

I think many people are not satisfied with that. They would like to leave behind a legacy, a name for themselves. Some people such as Alexander the Great and Julius Caesar did that through military means. Others, such as Leonardo da Vinci, did it through their art. Mathematical physics gives individuals the opportunity to do it by leaving behind their published work. If a mathematical physicist is fortunate, their theories live on long after they themselves have passed away. Notable examples are Newton, Maxwell, and of course, Einstein.

## Chapter 9

# How It All Started for Me

*The problem to address in cases of transformative choice is that, given the way the decision is naturally framed, you lack the ability to rationally determine the subjective values of the relevant outcomes, both because you cannot assign them at the outset and because the experience will change you in ways unknown.*

L. A. Paul [Paul (2014)]

Life is full of little incidents, each of which may have lifetime consequences. I never intended to be a mathematical physicist. At school I developed a passionate interest in astronomy, to the extent that when it came to deciding on a university degree, I chose *Astrophysics*.

There is nothing wrong with Astrophysics. Let me make that clear right now. It is a glorious subject, dealing with the origin, structure, and lifetimes of stars, galaxies, and the Universe. You can certainly have an excellent career in Astrophysics, one that will be rewarding and absorbing. If anything, there are more exciting developments going on and to come in Astrophysics than ever before, so it's a great time to start a career in that subject. You might even find yourself in space. Space science is booming right now and looks set to expand enormously.

At age seventeen, I set off by train to a distant land, Scotland, to start a degree in Astrophysics at the University of Edinburgh. That was the beginning of an exciting personal adventure that is still going on. Edinburgh is a city that I took to immediately. I found the people there and their culture one that I fitted in with ease.

During my first year, I attended lectures in classical astronomy, astrophysics, classical mechanics, and a whole slew of relevant topics in mathematics. The teaching staff included a solar astronomer, Dr Mary Brück, and an elderly mathematical physicist, Dr Robin Schlapp.

They (and many of the other staff) were inspirational and made lasting impressions on me. Dr. Brück did research in solar astronomy (that is, the study of the Sun), because when she started in astrophysics, the powers-that-be thought it was unseemly for a woman to be working at night on her own. That's what she told us in class. Her whole career was dictated by the archaic attitudes of others. How times have changed.

One of the topics she covered in her lectures was the *Equation of Time*. That models the discrepancy between *apparent* (what you actually see) solar time and *mean* (or clock) time. I always found that phrase, *Equation of Time*, fascinating. It suggests something deep and fundamental, whereas it's rather mundane and practical. It originated in antiquity, as astronomers even then recognized the difference between what they observed and what was theoretically expected.

There's a good joke involving time, clocks, and Edinburgh.

**On Time**

If you have ever been to Edinburgh, you'll know that at one o'clock in the afternoon, the *One o'Clock Gun* is fired from the Castle. It's a tradition dating from 1861, when a cannon was fired (from a different site) to allow mariners in the Firth of Forth to synchronize their clocks.

The story goes as follows. Old Jock had fired the cannon for forty years, and now he retired. The day after retirement, he went out to a local pub near the castle for a quick beer. He had never been to that particular establishment before, so found himself in the company of strangers.

Being a sociable sort of fellow, he soon struck up a conversation with another elderly gentleman, called Old Angus, standing at the bar. Old Jock asked him: "*And what did you do before you retired?*"

Old Angus replied. "*For forty years, I maintained the clock at St. Giles' Cathedral. I wound the mechanism and made sure it was running on time. So what did you do before you retired?*"

Old Jock smiled. "*That's a coincidence. For forty years I fired the Castle gun at one o'clock every day. I used your Cathedral clock chime to know when to fire the gun.*"

Old Angus choked. "*But I always adjusted my Cathedral clock to one o'clock when I heard your gun fire ...*"

If you ever become a mathematical physicist, you will encounter the importance of synchronizing clocks in Special Relativity. That has been the source of much interest ever since Einstein wrote his great paper on the subject in 1905 [Einstein (1905)].

Robin Schlapp was old, really old. At least, I thought so at the time. He was well over the standard retirement age of sixty five but that meant nothing. He was still working, his mind was bright, and he gave great lectures. He was a fine sight: a thin, tall, distinguished elderly man, going through heavy traffic on his bike to and from lectures, cycling *very very slowly*. I developed the theory that he never fell off because the laws of physics could not decide on which side he should fall over on.

There's a reason I have for remembering Doctor Schlapp. He was a kind and thoughtful lecturer and took the trouble to answer students' questions as fully as he could. Here's what happened when I had a question about Newton.

## Heresy

During his mechanics lectures to us First Years, Doctor Schlapp covered Newtonian orbit theory. That means using Newton's laws of motion and Newton's law of gravitation to show that planets move around the sun in elliptic orbits. When it comes to comets, the theory shows that some of them can move in hyperbolic orbits. Those are not periodic orbits: a comet on such a trajectory comes into the solar system once, goes round the sun once, and then goes out of the solar system never to return.

Doctor Schlapp analysed hyperbolic orbits using a set of mathematical functions called *hyperbolic functions*. These are related in a complex way to your normal trigonometrical functions (the sine and cosine functions).

I was greatly taken by his approach to the theory of comets, so much so that I looked up the hyperbolic functions in an encyclopedia (Google did not exist then). To my surprise, I discovered that they were devised in the 1760's by Giordano Ricatti and independently by Johann Lambert.

That really worried me. Newton published his great book, *The Principia*, in 1687 [Newton (1687)]. In it, he showed that his laws of motion and his law of gravitation predicted that comets moved on orbits shaped like so-called conic sections. Conic sections come in three basic types: ellipses, parabolas, and hyperbolas. The problem I had was that I found out that *The Principia* was published long before hyperbolic functions were invented. How did Newton work out hyperbolic orbits?

After the next lecture, I went to the front and asked the good Doctor about this date question. He nodded and said he would look into it.

True to his word, at the end of the next lecture, he handed me a copy of an old book. It was an account of Newton's life and had much historical detail in it that made a great impression on me. I can't recall what it said about hyperbolic orbits, but one incident about Newton was burned into my mind. It's about Newton's religious beliefs.

Newton lived in a very God-fearing age. In those days, you dared not say you didn't believe in God. Like the rest of that society, Universities were dominated by religious-minded people and anyone who did not toe the line in respect of established conditioning would not have a job there for long, to say the least.

You may know that Christianity of the Anglican variety (the one Newton was supposed to belong to) is associated with a belief in the Trinity. That's the idea that whilst there is one God, according to this belief, there are three *persons* (Christ being one of them).

Newton did not believe in the Trinity, but he was a Fellow of Trinity College, Cambridge. This was an ironic and uncomfortable state of affairs for him. He had to conceal his beliefs in order to retain his position at Cambridge, which he did apparently successfully.

According to the book Doctor Schlapp lent me, many years after Newton's death, an Anglican minister opened an old box in the attic of an old house and found some of Newton's hand written documents. When the minister read Newton's personal beliefs about the Trinity, he immediately shut the box and hid it, telling no one about Newton's heretical views for a long time. That was to protect the image of Newton as a great national figure.

Even long after his death, Newton's views about religion would have been harmful to his reputation.

Reading about Newton the freethinker made me realize that sometimes, it's wiser to keep some of our opinions to ourselves. This can apply also to mathematical physics. There have been occasions when I've found myself surrounded by theorists working in conjectural research areas such as *String Theory* and *Quantum Gravity*. I generally thought it wiser not to tell them that I thought they might be wasting their time.

## 9.1 Astronomy One

University is an amazing experience. If you get there, you will encounter people like Mary Brück and Robin Schlapp who will inspire you. And of course, you will enjoy the company of people just like yourself, all around you.

In those days, we students did not have the financial concerns of today's students. As my parents came from a pretty poor background, I had a student grant for maintenance, topped up by a modest parental contribution. The concept of University fees paid by students did not exist at that time. It was in a real sense the best of times.

I did not find the subjects a problem, except for one. In the first term of our first year, all six of us budding astrophysicists (it is a specialist sort of subject, so the class was really small) *failed* the Astronomy One Christmas Class exam. All six of us. I can still see the look of horror on the face of Prof Brück, the distinguished head of the Astrophysics department, when he met us subsequently to discuss this disaster.

I don't want to minimize the seriousness of this incident. It was unprecedented, as far as I know and it was certainly not the fault of the staff. I think it was a statistical fluke, that all six of us were just getting used to student life and perhaps needed a jolt. In my case, it was the focus on *spherical astronomy* (geometry on a sphere) in that first term that did me in. I found it a bit boring. I wanted to get onto more exciting things, such as spiral galaxies and expanding universe.

But, as they say, you have to learn to walk before you can run. That is a truth that you will realize the hard way, and is part of the deal. Absolutely every job, occupation, activity, enterprise, and suchlike, will require you to go through boring bits. You just have to put up with them. Hopefully, the boring bits (which are usually at the start) will not last for ever. With some jobs, the good bits far outweigh the inevitable boring bits.

What were the consequences of the failed exam? None of great significance really, as far as the rest of the Universe was concerned. But it had a lasting impact on me, as I will explain in a moment. As for my fellow students, it was just another one of those things. Subsequently, at least one of my five colleagues went on to do great things in Astrophysics, so it was not a career-threatening incident. Christmas class exams are there precisely to give you a reality check before you do the end of year exams that do count.

I recount this episode in my life to reassure any person who has failed an exam. It does not mean that you are a failure. That will depend on what you decide to do about it.

## 9.2 The Handbook

After the Christmas break back home, there I was, sitting in my student room back at Edinburgh in the New Year, reflecting on my results and feeling a bit miserable. I was not used to failing any exams.

By some chance, I had brought with me back to Edinburgh the University Admissions Handbook that outlined all the various degree courses that the University had to offer. Of course, I had read its section on Astrophysics a year earlier, back home, when I was deciding on which University to go to do Astrophysics. Rather absent-mindedly, I picked up the Handbook and opened it. As I turned its pages at random and without being consciously aware of what I was doing, my attention was suddenly taken by the page on the *Mathematical Physics* degree. It was written by a certain Professor Kemmer, of whom I had never heard, and this turned out to be one of those decisive moments in my life that changed me forever. Have you ever had one of those moments?

What Prof Kemmer wrote was direct and inspirational. Before I read his words, I had only a vague idea of what mathematical physics was. I had thought it was half mathematics and half physics. No, that would not be for me. I was not interested in doing things by halves. I simply had never thought about it seriously, until I read what he had written.

The first thing Prof Kemmer made clear was that mathematical physics is emphatically *not* a made-up subject, a hybrid if you will. It is not a Joint-Honours course that you can get at some Universities, such as Joint-Honours Mathematics-and-French, or Joint Honours Physics-and-Philosophy.

**Joint Honours**

There is absolutely nothing disreputable or unacceptable in doing a Joint Honours degree. Let me be quite clear about that. They can be entirely right for some people. Take the case of Joint Honours Physics-and-French. Someone who likes Physics and thinks that they would like to work as a physicist in France might well decide that that was the degree for them.

Years later, I found myself involved in running some Joint Honours courses in the maths department where I worked. I used to imagine inventing some far-out joint-honours courses. How about Joint Honours Drama-and-Bricklaying? That could be useful to an out of work actor getting a job on a building site whilst they waited for auditions. Or Joint Honours Catering-and-Forensic Science? That could be useful to someone who had a bad meal at a restaurant.

Mathematical physics is a single, thoroughly unified, beautiful subject. *It is the mathematics of physics.* It is what Newton did, what Einstein did, what Stephen Hawking did. It has a well-established, international community of thousands of like-minded individuals, young and old, with its own journals, books, University departments, University degree courses, and conferences (meetings of like-minded people).

*You* can join it. It's not an exclusive club, unless you exclude yourself. It's a *glorious* subject that should interest anyone with half an interest in the universe in which they live. It brings together all those aspects of intellect that make us humans what we are: *thinking machines.*

What Prof. Kemmer wrote in essence was that

**Mathematical Physics is the study of the physical universe using any of the tools of mathematics that are required.**

Those aren't his exact words, because I lost that handbook decades ago, but that's essentially what he meant and it's the way that I think of it. It's the way that I think it should be defined to be. I will have more to say about this point in other chapters, because some mathematical physicists seem to have forgotten the 'physical universe' bit of my definition.

As I read more of Prof Kemmer's description of the Mathematical Physics degree on offer, I had some sort of epiphany, a sudden realization that *this* was the subject I really wanted to do. Suddenly, *Astrophysics* seemed to flow away from my focus of attention and was replaced forever by *Mathematical Physics*.

Don't get me wrong. It's not that I suddenly disliked Astrophysics. It's simply that I now I was absolutely sure I wanted to do Mathematical Physics. I went to my personal tutor and he arranged for me to switch to mathematical physics in my second year.

I sat the Astronomy One exam in May and passed it. That was even acknowledged on my Mathematical Physics degree certificate when I graduated three years later. The question I sometimes ask myself is: what would have been the course of my life if I had passed that exam at Christmas? Looking back at it all those years, I'm grateful that I failed it.

## 9.3 Overview

If you've read my anecdote *The Spinning Floor* in Chapter 1, you'll know that I haven't hesitated to change the direction of my life when I've believed it was important. It's not always easy, but speaking from personal experience, I would say that jumping into the unknown and taking a risk has paid off for me on more occasions than it has disadvantaged me.

If you do contemplate any serious change of direction in your life, my best advice is to research all options thoroughly and try to find positive reasons for the change. But sometimes, you cannot rationalize your decision. In cases like that, you may have to trust to instinct and take the plunge with your eyes closed [Paul (2014)].

# Chapter 10

# No Good at Maths?

*Philosophy is written in this grand book, the universe, which stands continually open to our gaze. But the book cannot be understood unless one first learns to comprehend the language and read the letters in which it is composed. It is written in the language of mathematics, and its characters are triangles, circles, and other geometric figures without which it is humanly impossible to understand a single word of it; without these, one wanders about in a dark labyrinth.*

Galileo, *The Assayer* [Galilei (1623)]

We come now to the hard part. You may be **no good at maths**.

Or so you have been told, or even believe about yourself. Nonsense. I'll prove it to you. If you can read, you can do maths of enormous depth and value, as I'll show in this chapter.

There is some mythology concerning Einstein and mathematics. Some people say he was bad at maths and failed maths exams. The truth is far more interesting. He was very good at maths. He was not, however, principally interested in maths, but only in the application of maths to his physics theories. In [Isaacson (2007)] you can read a good account of how he realized almost too late in the day that mathematics is an important ingredient in mathematical physics.

Whenever I taught mathematics to First Year physics students, I always emphasized the idea that *a physicist who develops their mathematical skills becomes a better physicist.*

## 10.1 Sanitized maths

For many years, I never understood why maths was taught in the way it was taught to me and probably is still taught in many schools around the world. One of my first memories in secondary school (from about age 11 in the United Kingdom) was going over a standard textbook on Euclidean geometry. It's a fabulous subject, discussing circles, triangles, parallel lines, Pythagoras' Theorem, and so on.

One point constantly nagged at me at the time, a point that was not clarified until many years later, when I started to work in an actual department of mathematics. I could not believe that mathematics consisted of Theorem 1, followed by Proof of Theorem 1, followed by Theorem 2, followed by Proof of Theorem 2, and so on, all the way to the end. Why? How? Who?

On top of that, maths had examinations, which for the record, I was not bad at (until my passion for physics got the better of me). I never understood the need for speed in examinations. What does it prove to answer four questions out of a choice of six in two hours?

As I said, many years later, all became clear. Real mathematics does not consist of an endless string of theorems and proofs. Real mathematicians take ages to solve their problems. Real mathematicians do not time themselves to solve their problems in two hours. They can take years to understand a problem and construct a theorem. They make mistakes. They argue about infinity, the infinitesimals, probability, and base much of their work on educated *guesses* called *conjectures*.

### Mathematicians are Human

I remember years ago reading a one page article in *Scientific American* about *mathematical conjectures*. A conjecture is just an educated guess. They are dangerous beasts indeed, because they can be wrong. The article examined the risks taken by mathematicians when they decide to use a conjecture in their research.

Imagine a group of sophisticated and advanced mathematicians stuck at some point in their mathematics research. Despite their best efforts, they cannot clear a conceptual road-block that is blocking all future progress in their field. Suppose further that they can see that the road-block would be removed if just one particular idea were true. That idea, when examined, seems to work in all cases examined by the mathematicians. It looks, sounds, and intuitively feels, completely reasonable and true. Suppose the mathematicians decided to *assume* that the idea was true. That would allow them to continue working past the road-block deeper into their research program. They imagine that one day, someone will prove rigorously that the idea is true after all.

The danger for those mathematicians is that the idea, known as a conjecture, might be false after all. It would not take much. In mathematics, just one counter-example of a proposed theorem would be enough to destroy that theorem's value to mathematicians. If that happened, any work done past that road-block would be worthless. All the years of work done subsequently that was based on the validity of that conjecture would collapse like a house of cards in a hurricane.

The article I read discussed the sad case of a long standing school (group of professionals) of mathematicians researching into an exotic branch of mathematics called *algebraic geometry*. What happened with them is that the careless use of conjecture and inadequately rigorous arguments led to the collapse and demise of an entire line of research done over many years, when other mathematicians proved that their conjectures were false.

If and when mathematicians reach some sort of understanding of what may have taken years to sort out, then they sanitize everything in their published accounts of their work. They tidy up their equations. They give names to theorems. They write out neat proofs. Then they write papers and text books starting with Theorem 1, followed by Proof of Theorem 1, followed by Theorem 2, followed by Proof of Theorem 2, and so on, as if that had all come to them in a flash of inspiration.

All of that sanitizing is done for public consumption. Real mathematics is never done like that. Real mathematics is done in offices, in tea rooms, on buses, on long country walks, fierce discussions between colleagues, endless rewrites of papers, and all the countless other activities associated with any human activity. My first impression of mathematics, therefore, was of no more than a sanitized, tidied up, textbook version of hard work that had taken many generations of mathematicians centuries or even thousands of years to stumble upon and develop. No one told me at the time.

## 10.2 The Shores of Infinity

I want at this point to lead you through a straightforward but illuminating calculation. The instructions I give are basic and should be understandable by most people. Regrettably, too many people, young and old, think that they *can't* do maths, when in fact they *could*, if they were given just a bit of encouragement to believe in themselves as mathematicians.

Read the following and aim to understand each small step. If you do that carefully, you will end up with an equation that takes you to the other side of the mathematical universe, using basic arithmetic. That should convince you that you *can* do maths, if you give yourself half a chance.

You will agree, I hope, that

*one plus one is two.*

Before I go on, let me introduce some *notation*. Words are generally useful, but they can become inefficient when it comes to mathematical concepts. We may well understand the idea of *one plus one is two*, but it's tedious to write that expression out in full every time we use it. It's easy to see why. In English, I need nineteen symbols, including spaces, to type it out. I wonder how many symbols it would take in your language? It would help if we could express such ideas more economically.

What maths does all the time is encode such ideas symbolically, that is, using symbols instead of words. That doesn't change the meaning of those ideas. An idea in mathematics is usually expressed using marks on paper that are different to letters. There's nothing mysterious about that.

Any symbol is just a mark on a piece of paper. This of course includes letters, which are also symbols. The point is, a symbol makes anyone looking at it think about the agreed meaning associated with that symbol. That's of course how reading and writing work. It's that principle that allows you to understand what I have written, even if I am long dead (probably true if you're reading this in a second-hand book store). Mathematical symbols are exactly the same thing.

One of the big problems with maths is that if you are unused to the symbols, then you can't understand what they represent, and that's often why maths looks incomprehensible to the person who doesn't do it for a living.

There are, as we know, many hundreds of languages on this planet. English is one of them. So is Chinese, Urdu, Finnish, and so on. Each language has its own particular set of letter symbols, words, and grammatical rules. That makes for a great deal of confusion between nations and job employment opportunities for translators.

With maths, it's different. Mathematicians all over this planet have decided to use the same consistent set of symbols to express mathematical ideas, regardless of the nationality of any mathematician.

### MCMLXXXIV

The universality of mathematical notation is a fairly recent development. Take the ancient Romans. It's well-known that they had a pretty clumsy way of expressing their numbers. For example, the date *1984* would be expressed in Roman numerals as $MCMLXXXIV$. I need four symbols to write *1984* and nine to write $MCMLXXXIV$, so just by counting symbols, Roman numbers are in this example at least twice as inefficient as our ordinary decimal notation, 1984. Not only that, but doing maths using Roman numerals, such as multiplying two numbers out, gets complicated. And please don't ask about division in Roman.

Getting back to the point, it has been generally accepted by mathematicians all over this planet that the statement *one plus one is two* can be represented by the symbols

$$1 + 1 = 2. \tag{1}$$

Notice that I've labelled this equation by (1), because we will come back to it presently.

This world-wide agreement amongst mathematicians is astonishing, given all the problems between different nations and cultures. Mathematics and its notation does not "belong" to any particular culture. It's not English or Japanese. It belongs to everyone, including me and you. This means that people all over the planet can join the club and be mathematicians with the same rules of the game. The universality of mathematics underlines the fact that we are all living on a single planet and that we have many things in common, particularly in the way that we can think.

## Is Mathematics a Language?

As I wrote out the paragraph immediately above, I started to think about the universality of mathematics (meaning how it belongs to all humanity).

One curious thought came to my attention: that mathematics is not a "language" such as those you learn before you go on holiday. When you learn a new language, you have to learn the meaning of the words, you have to learn the alphabet (or its equivalent) if that alphabet is unfamiliar to you, you have to learn how to pronounce those words, and how to string them into proper grammatically correct sentences.

With maths it's different. When you see the symbols $1 + 1 = 2$, you read them out *in your own language*. So an English person would say "*one plus one is two*", whereas a French person would say "*un plus un est deux*". Each person would understand in the same way the same mathematical symbols. There's no actual language called *Mathematics*, with its particular rules of pronunciation. On reflection, therefore, I would not say that mathematics was a language, *per se*.

This raises the obvious question: *what is mathematics*?

The best answer I can come up with is that mathematical notation is a *symbolic representation of certain universal ideas.* Since those ideas are common to all humans (or should be), then mathematics is universal in that respect.

Why do so many people, including perhaps you, think that maths is an alien discipline?

The best answer I can come up with is that anything that has lots of bits to it (such as a computer program or a music score) can seem incomprehensible at a beginner.

How do you get around that? The golden rule here is: *don't let mathematics fool you.* If you *modularize* any equation (meaning you look at each bit separately and carefully), then you can begin to reduce the apparent complexity of any piece of mathematics. Certainly, it may take a lot of time in some cases, but then, mathematicians who do it for a living take a lot of time over it as well.

## The Many and the Small

Mathematics is not a divisive subject, usually, but I have to be honest. There are one or two branches of mathematics that have had some conflicts of opinion that were occasionally intense and unpleasant. Two of these are *Statistics* (the theory of probability applied to real life situations) and *Infinitesimals* (the theory of the infinitely small). In such cases, the dispute was/is about the validity of the principles being assumed, not about the symbols used.

Let's get on with our calculation. If you know about *one plus one*, you may also know that *two halves make a whole*. That idea is expressed in maths by

$$\frac{1}{2} + \frac{1}{2} = 1. \tag{2}$$

The great trick in maths is to take bits of it from one place and use it in another place. The first thing to do is to rearrange equations (1) and (2) as follows:

Our first equation (1), $1 + 1 = 2$, becomes

$$2 = 1 + 1 \tag{3}$$

and our second equation, (2), $\frac{1}{2} + \frac{1}{2} = 1$, becomes

$$1 = \frac{1}{2} + \frac{1}{2}. \tag{4}$$

All I've done here is to swap over the left-hand side and the right-hand side in each equation.

The next step is to use equation (4) in equation (3) as follows:

$$2 = 1 + \underbrace{1}_{\frac{1}{2}+\frac{1}{2}},$$

giving

$$2 = 1 + \frac{1}{2} + \frac{1}{2}. \tag{5}$$

Got it? Right, we can carry on. The idea now is to keep on doing exactly the same thing, over and over again, replacing the last term each time with an arithmetically equivalent but different looking expression. For instance, we should know that

$$\frac{1}{2} = \frac{1}{4} + \frac{1}{4},$$

so we can rewrite (5) as

$$2 = 1 + \frac{1}{2} + \underbrace{\frac{1}{2}}_{\frac{1}{4}+\frac{1}{4}},$$

giving

$$2 = 1 + \frac{1}{2} + \frac{1}{4} + \frac{1}{4}. \tag{6}$$

Assuming you've followed so far, we just keep going. Hopefully, you can work out for yourself that before long, we would have in front of us on paper

$$2 = 1 + \frac{1}{2} + \frac{1}{4} + \frac{1}{8} + \frac{1}{16} + \frac{1}{32} + \dots, \tag{7}$$

where the three dots ... indicate that *we could go on for as long as we liked.*

Those last three dots in equation (7) represent one of the significant ideas in mathematics: the concept of *infinity*. The key question here is: *where do we stop?*

Leave that question aside for the moment. We've hit a bit of a snag. The symbols are beginning to pile up on the line. If we kept on going, we would start to cover the page. Then the next few pages. Then the whole book, and so on. Do you see the problem? Equation (7) is a bit much to write out as it stands. So mathematicians have invented the notation

$$\frac{1}{2} + \frac{1}{4} + \frac{1}{8} + \frac{1}{16} + \frac{1}{32} + \dots \equiv \boxed{\sum_{n=1}^{\infty} \frac{1}{2^n}}$$

to express everything about equation (7). First, the big symbol $\Sigma$ means "add up". Next, the symbol $n$ labels a typical element or term in the sum, starting with $n$ equals one (the $n = 1$ bit) all the way up to infinity, denoted by the $\infty$ symbol. What follows the big symbol $\Sigma$, the $\frac{1}{2^n}$ bit, represents the typical term in the sum. For example, $\dfrac{1}{2^3} = 1/(2 \times 2 \times 2) = 1/8$, and so on.

I asked: *where do we stop?* Actually, we could stop at any point and have an exact statement. By this, I mean equations (3), (5), and (6) are each mathematically true. I'm going to choose now to go on forever. *To infinity and beyond!*

It's the infinity symbol, $\infty$, that's the big deal. The concept of infinity lies at the borderline between the possible and the impossible, between reason and faith, between the observable and the unobservable, between the definite and the indefinite, and between proof and conjecture. It's the difference between fact and fiction, if you like, because I could stop somewhere and have an exact sum, or else I could go on forever (or so I imagine) and still get the same exact sum. The notation requires you the reader to understand what is going on. The symbol $\infty$ is not normally thought of as an actual number (although in some areas of maths it is treated as if it were). Rather, it's more like a *process*, telling us to keep going without stopping or reaching any limit.

There's a great deal of history and controversy in mathematics associated with infinity. Let me say that some mathematicians like it, others avoid it. Look up online the controversy between the mathematicians Kronecker and Cantor. When you see what they thought of each other, you will realize that any particular problems you may have in maths are nothing in comparison. By the way, equally intense conflicts of opinion have arisen in physics. Look online for what Heisenberg and Schrödinger thought about each other's theories in quantum mechanics. *Rude* would be a mild adjective.

Returning to our calculation, subtracting 1 from both sides of equation (7) gives the final amazing result:

$$\boxed{1 = \sum_{n=1}^{\infty} \frac{1}{2^n}}.$$

If you thought you could not do advanced mathematics, but have followed what we've just covered, then there you are, sunbathing on the shores of infinity, wearing no more than basic arithmetic. If you can do this, what else could you do?

## 10.3 Practice makes perfect

I'm not going to pretend that you could do advanced mathematics just like that. But then, you wouldn't watch a game of professional football and say "I could do that!" without expecting to do a whole lot of training and practice, either. You would naturally expect to train and practice a lot, way before you could expect to be taken seriously as a footballer. It's the same in maths. That's where you achieve your ambition: practice, practice, practice.

The key point about mathematics is to do it step by step, slowly and carefully. Bit by bit you get used to each new idea, each new trick of the trade. It may be hard in places, but if you take it easy and give yourself enough time, you should be able to climb up most mathematical hills.

There's four points I want to make here. I hope they're helpful.

### *Linearity*

The first point is that the study and learning of mathematics is rather different to what happens in many other subjects. Take *History*. I've nothing against History. It's an important subject in helping us find our place in the Universe, where we came from, how our societies developed, and giving possible lessons for planning the future. But you don't need to study all of it. You could decide, for instance, to study only Twentieth Century history. Or perhaps only the ancient history of your own country. With maths it's different. You would find it hard if not impossible to study real numbers (the *reals* as they're called in the trade) if you had not studied arithmetic first. And you would not understand arithmetic if you hadn't got used to the integers (the counting numbers such as *one*, *two*, and so on) first.

I worked in a department of mathematics for over thirty years, and on occasion heard maths described as a *linear* subject. That's like someone climbing up a rope. To get to the top of the rope you have to start at the bottom and pass through every other bit of the rope in proper order before you could get to the top. That's not the case with History.

### *Individuality*

This second point is important. Everybody is different. Even in a department of mathematics, you'll find that almost no one knows everything. Mathematicians tend to specialize. Some of them love subjects like linear algebra and group theory and hate subjects like functional analysis and partial differential equations. Or vice-versa. In a department of mathematicians, you'll find people who are statisticians (averages and so on applied to real life situations) mixing with number theorists (people who explore the properties of numbers as an abstract discipline). Here's an anecdote from real life illustrating this point.

### Stats and Maths

In the maths department where I worked for many years, we used to have regular departmental meetings. They were held usually three or four times a year.

On this occasion, all of us staff were seated around the edges of a big room, backs to the walls, all looking inwards, with a big empty space in the middle. It was a bit like King Arthur and His Knights of the Round Table (only our table was square). On one side sat *Head of Department* and the other professors. We came to an important item on the agenda: what to do about the teaching of mechanics to our second year mathematics students?

You might think mechanics has nothing to do with maths. You would be **so** wrong. Indeed, much of modern mathematics was developed *because* of a need to understand mechanics. The mathematics of mechanics is an extraordinarily beautiful and central part of mathematics, so much so that some people have confused the two and think mechanics **is** only mathematics. I'll have more to say about that and those people later on in this book.

Anyway, back to the discussion about second-year mechanics. At one stage, *Professor Statistics* volunteered the statement that he thought that "*statistics students don't need to know mechanics*".

*Head of Department* immediately responded by saying that **he** "*never thought that statistics was a branch of mathematics in the first place.*"

Have you ever seen a volcanic eruption? There was an excellent representation of one coming from *Professor Statistics.*

The moral here is that almost certainly you will find that you have definite preferences in maths. Some branches of maths will appeal to you whilst others will leave you cold. I've often thought that people who think they can't do maths do so only because they haven't yet found some branch of it that turns them on.

It's that diversity in mathematics that allowed me, a mathematical physicist at heart, to work in a maths department.

### *No universal toolkit*

The third point is that I don't know of any single set of ground rules for solving mathematical problems, except for having patience, taking time, thinking a lot, and being prepared for going wrong before stumbling on a solution to some problem. The following anecdote illustrates the point.

#### Rock Pools

Have you ever taken off your shoes and socks and walked in the surf along a sandy beach on a glorious sunny day? It's one of the great natural (and free) pleasures of life.

What I like about it is that, if there's no one else around, you can close your eyes and just walk in the surf. You don't have to have a sophisticated plan or strategy. You first put out your left leg and step forwards with your left foot. Then you do the same with your right foot. Then your left. And so on. You can even do it with your eyes closed, feeling the water swirling around your feet, splish-splashing along, absorbing the sounds and sensations. You can keep repeating this all along the sandy part of the beach.

Another analogy is driving along a long, straight, deserted road in Australia. Put your car onto cruise-control and take a nap for ten minutes. On second thoughts, perhaps not. You may hit a kangaroo.

Maths is not like that. Maths is much more like trying to cross the *rock pools* at the end of the beach. There you would be well advised to **not** close your eyes. Moreover, repeating your steps would almost certainly be dangerous. On a rocky part of the beach, you had better treat each little ledge, each step, each corner, each pool, with the care it deserves. The little trick that you used to leap over that last rock pool just won't work when you have to climb over the huge rock in front of you. A metre jump here may do, but over there, you would fall short and probably break a leg.

So it is in maths. Each problem has its own peculiarities, its own charms, its own difficulties, its own impossibilities. Hopefully, your problem will have at least one way to solve it. But you can't be sure. If you've ever heard of *Fermat's Last Theorem*, you'll know that it's relatively easy to state, but it took hundreds of years to prove [Singh (2005)].

This third point is what gives many people a lot of pleasure solving maths problems. Maths need not be boring. It can make you just lie down and think about a particular problem, exploring mental pathways that perhaps no one else in history has ever passed through. It's not impossible that *you* will stumble on some proof of a previously unsolved problem that has defied the best attentions of mathematicians for centuries.

Having said that, you should be careful. Professional mathematicians all over the world still often receive from amateur mathematicians false "solutions" to famous long-standing "problems", such as "*squaring a circle*", or "*proving Euclid's fifth postulate in Euclidean geometry*". These are not just unsolved problems: they are *unsolvable*. Mathematicians can *prove* that they cannot be solved. You can find these proofs readily on the internet.

### *Rely on mathematics*

Mathematics has one unique characteristic: *if you do a calculation correctly, you can stake your life on it.* No other discipline that I know of can guarantee you that, not even mathematical physics, because mathematical physics is based on empirical evidence, which is invariably finite and therefore incomplete.

I suspect, after dealing with mathematics students for many years, that the subject attracts students in far greater numbers than mathematical physics does *because it is safe.* I even overheard a student tell another student that that was why they had chosen to do mathematics. A mathematician can choose their axioms and their postulates, and then work out rigorously all the possible theorems associated with those axioms and postulates.

However, safety isn't everything. I think physics is harder than mathematics precisely because physics is never safe. New experiments may confound the theories you've been using all your life. On the other hand, Euclidean geometry remains a rigorous and useful branch of mathematics, after two thousand years. It's only called into question when people try to apply it to the empirical universe.

Perhaps you can understand now why I think mathematical physics is harder than mathematics or physics. Mathematical physics lives right on the borderline between the total security of mathematics and the empirical uncertainties of physics. You use the certainty of mathematics to model unproven intuition about physics. Sometimes I think of mathematical physics as the ultimate art form.

## 10.4 Sell by date

Mathematics is the only discipline which, if it's done properly, never dates. What was done by Euclid and other ancient mathematicians still stands today. Physics, on the other hand, is based on doing your best to undermine everything you have ever believed in. The laws of physics are not absolutes. They are inferences based on finite information, and should be tested constantly. That's why I admire the experimentalists like Maris and Afshar in their attempts to undermine conventional ideas in physics. Maris attempted to "split" electron wavefunctions, hoping to see fractional electron charge [Maris (2000)]. Afshar attempted to circumvent Heisenberg's Uncertainty principle [Afshar (2005)]. It's that sort of experimentation that simply has to be done and supported. It's contrary to *Nullius in Verba* for any mathematical physicist to demean such experimentalists.

As more information comes in, those inferences may change. Once, physicists believed in the validity of classical mechanics. Nowadays they generally think of it as a pretty good approximation to a more correct view of reality based on quantum mechanics. In physics, theories invariably have finite shelf lives.

These differences between mathematics and physics hits mathematical physicists pretty directly. The longevity of mathematics is no defence against the volatility of physics. That means that mathematical physicists cannot expect their theories to survive beyond say a couple of hundred years. We would be foolish, for example, to ignore quantum mechanics and keep using classical mechanics, unless we restricted our attention to narrow domains of applicability. For instance, classical mechanics would be good enough to model river flow, but no good in quantum optics.

One of the great issues in mathematical physics is: *how long will quantum mechanics survive?* Perhaps you will be the mathematical physicist who finds a better theory.

# Chapter 11

# Einstein's Brain

*No one can claim to have explained Einstein's genius. For all we know, a person with big inferior parietal lobules could just as easily have become a great home builder or billiards shark as the man who changed our conception of the universe. For Einstein did more than manipulate mental images. He sought and found images that captured the fundamental aspects of physical reality, and converted them into appropriate mathematical equations and empirical predictions. These gifts surely lie in the microcircuitry formed by trillions of synapses in many parts of the brain, and we are not going to work out that wiring diagram in Einstein or anyone else any time soon.*

Steven Pinker [Pinker (2015)]

Even before he died, some people assumed that Einstein's brain had some special features that made him exceptional. It's not surprising, therefore, that just after he died in 1955, his brain was removed and has been studied and discussed ever since, amidst controversy.

You can read about what happened online. The conclusions reached by early investigators have been disputed [Hughes (2014)]. One of the points made by critics of the idea that his brain was special is that *every* brain is special, in that no two human brains are exactly alike.

It's true that there are regions and structures in every brain that have common features and functionality, such as the *hippocampus*, which is involved in memory. Beyond that, however, there are immense differences between any two brains, because of the fundamentally critical matter of how the various parts of the brain are *connected*. How our brains get connected is a fascinating and important story in its own right, involving experience, short term memory, long term memory, and sleep [Walker (2017)].

## 11.1 Brain myths

### *Fixed number myth*

Years ago, it was asserted and widely believed by neuroscientists that each brain has an ever decreasing store of *neurons*, which are the fundamental cells involved in the transmission of information throughout the brain and other parts of the body[1]. According to this myth, one of the hazards of alcohol and drug addiction or brain injury would be a loss in the overall number of neurons.

There is now strong evidence for *neurogenesis* (the generation of new neurons) in adults [Wade (1999)].

### *The ten percent myth*

Another popular myth is that only ten per cent of anyone's brain is normally used. In consequence, it is often asserted by the ignorant that with the correct training and stimulation, ordinary individuals might be able to access the remaining power of their brains and thereby achieve great things, comparable to those of Einstein. The plain fact is, the brain is an immensely complex structure that has to work intensively at all times in order to function properly [Boyd (2008)]. Surprisingly, our brains do more work when we are asleep than when we are awake, contrary to our naive intuition [Walker (2017)].

### *Adult wiring is fixed*

Another myth is that children and young adults can be influenced to think (conditioned) in various ways, but beyond a certain age (twenty-five to thirty), that is not possible.

Fortunately, that's not true. There is empirical evidence to support the assertion that our brains are being constantly rewired by our experiences, even as adults [Giang (2015)]. If such rewiring did not occur, no one would ever "change their mind" or opinions about beliefs that they had held for years. Certainly, politicians hoping to win elections have always assumed it's possible rewire voters.

### *Dyslexic Einstein?*

A common, popular belief is that Einstein was dyslexic. He *may* have been, but it's not conclusive [Eide (2014)].

By any yardstick of social interaction, Einstein was extraordinarily normal. He was attractive to women and had life-long friends. He communicated with many people scientifically at conferences and in writing, and there is no evidence that he was difficult to deal with in scientific matters. It's true that he was in general a lone worker, the archetypical mathematical physicist like Newton, Maxwell, Schwinger, Emmy Noether, and Dirac, each of whom published great research by themselves.

However, it should be understood that all of these "lone-wolf" theorists communicated constantly with wide circles of theorists and were fully aware of current empirical research when they created their mono-authored works of genius.

## 11.2 Einstein's view about education

If we want to model ourselves on Einstein, it's reasonable to understand the way he thought about matters such as education. He was a brilliant, creative thinker. Did he actually "know" a lot of facts? What were his views about rote learning (learning masses of technical facts)? Here's a revealing anecdote that may help teachers and students appreciate the difference between *learning* and *understanding*.

### Edison's Questionnaire

The historical evidence about Einstein's views on college education can be found in a book published in 1947, eight years before Einstein died [Frank (1947)].

When Einstein came to the United States in 1921, he was internationally famous and something of a celebrity. The great inventor Thomas Edison was still alive and holding strong views about the value of college education. According to Edison, education should be directed towards relevant facts. Edison went so far as to prepare a questionnaire designed to test people's practical knowledge. He believed that college students would not in general be able to answer his questions.

When Einstein was give a copy of this questionnaire, his answer to the question *What is the speed of sound*? was that *He didn't know. He did not burden his memory with facts that he could easily look up in any textbook.*

Edison thought that college education was useless. He had attended school for only a few months and never went to college. Despite that, he became a successful self-made entrepreneur and inventor. His background and life experiences had clearly conditioned him to look down on formal education.

Einstein, who was as creative a theorist as Edison was as an inventor of practical devices such as light bulbs, disagreed with Edison's view on education, remarking that "*It is not so very important for a person to learn facts. For that he does not really need a college. He can learn them from books. The value of an education in a liberal arts college is not the learning of many facts but the training of the mind to think something that cannot be learned from textbooks.*"

### 11.3 Einstein and music

When I was much younger, I was convinced that science was paramount and everything else, such as religion, politics, literature, art, music, theatre, and so on, were useless activities and could be (and should be) cut out of a person's life. The fact that I enjoyed playing the piano was an inconsistency I didn't want to resolve.

Over the years, I changed my mind. In other words, my pro-science/anti-arts conditioning changed from certainty to doubt. One factor that contributed to this change was seeing the logical Science Officer *Mr Spock* in the Star Trek Original Series episode *The Way to Eden* playing a lyre in a jam session with some space hippies. Later, I found out that Einstein played the piano and the violin (which he called *Lina* and by some accounts played well). Why would he waste his precious thinking time trying to scrape lengths of horsehair against four strings of catgut in order to produce a noise that, in a beginner, sounds like a cat with tooth-ache?

The answer lies in the way the human brain works. We are not robots, despite whatever *Sheldon Cooper* of *The Big Bang Theory* believes. A mountain of neuroscientific evidence points to sleep as a critical component in creativity [Walker (2017)]. If that were all, however, then having a good night's sleep should produce great thoughts during the day as we sit at our desks.

In my experience, having *any* new thoughts, let alone creative ones, gets more and more unlikely the longer I sit at my desk, even after a good night's sleep. I need to take breaks. By experience, getting up from my desk and taking a short walk, or even extemporizing on my piano, every hour or so works wonders. Another good way to revitalize a jaded brain is to do something apparently pointless and time-wasting, such as watching a good film or listening to some music (I don't advocate listening to music *during* work at a desk, but it may work for you).

You can find informative accounts of Einstein's musical ability online. For example, according to his second wife Elsa,

> *...he played Mozart so beautifully on the violin. He also plays the piano. Music helps him when he is thinking about his theories. He goes to his study, comes back, strikes a few chords on the piano, jots something down, returns to his study.*
>
> [Head (2015)]

## 11.4 Einstein and religion

The fact that Einstein was agnostic (that is, not religious in a formal way) is not evidence that scientists should be atheists. Michael Faraday was arguably the greatest experimental physicists of the Nineteenth century yet he was a *Sandemanian*, a now extinct branch of Christianity. The point to note here is that neither Einstein nor Faraday used or referred to their views about religion in their scientific work.

My interpretation of these historical facts is that there should be no conflict between being a scientist and having views about religion, one way or the other, *provided* you take care not to mix the two, or claim they are different manifestations of some "truth". They are not. Don't write a scientific paper and claim it "proves" anything about religion. Likewise, don't assert that your religion has anything to say about empirical science. Religion does not respect *Nullius in Verba* and that's all I need to know. As Jonathan Swift observed: "*...Reasoning well never make a Man correct an ill Opinion, which by Reasoning he never acquired*" [Swift (1721)].

## 11.5 Einstein and mathematics

If you want to emulate Einstein but are worried about mathematics, then here's a paragraph I found from the *New Yorker* magazine that may help you.

**It's Worse for Me**

On January 3, 1943, a junior-high-school student named Barbara Lee Wilson wrote to him [Einstein] for advice. "*Most of the girls in my room have heroes which they write fan mail to,*" she began. "*You + my uncle who is in the Coast Guard are my heroes.*"

Wilson told Einstein that she was anxious about her performance in math class: "*I have to work longer in it than most of my friends. I worry (perhaps too much).*"

Four days later, Einstein sent her a reply. "*Until now I never dreamed to be something like a hero,*" he wrote. "*But since you have given me the nomination I feel that I am one.*"

As for Wilson's academic concerns? "*Do not worry about your difficulties in mathematics,*" he told her. "*I can assure you that mine are still greater.*"

[Strogatz (2015)]

Einstein was not being modest in his reply to Barbara. He was certainly not the best mathematician on the planet, and he did no great experiments in physics himself (that I know of). Whether or not you do learn to think like Einstein will depend on quite a lot of things, such as whether you *want* to, for a start. Another will depend how much effort you put into it. And, most likely, what sort of conditioning you have already received, because too much of that may already have spoilt prospects in that respect.

## 11.6 Einstein and intuition

There are few clues about how great mathematical physicists actually achieved their results. In the case of Einstein, there are suggestions that his sense of *visualization* was strong. By this is meant the ability to see principles of physics re-enacted visually (in the mind), rather than being worked out from mathematical manipulation of symbols on paper.

At age 16, Einstein went for one year to the Argovian cantonal school in the town of Aarau in Switzerland, and this may have been a critical experience regarding his subsequent achievements. The school was run on the educational principles of Johann Pestalozzi, which suited Einstein perfectly:

*It was a perfect school for Einstein. The teaching was based on the philosophy of a Swiss educational reformer of the early nineteenth century, Johann Heinrich Pestalozzi, who believed in encouraging students to visualize images. He also thought it important to nurture the "inner dignity" and individuality of each child. Students should be allowed to reach their own conclusions, Pestalozzi preached, by using a series of steps that began with hands-on observations and then proceeded to intuitions, conceptual thinking, and visual imagery. It was even possible to learn-and truly understand- the laws of math and physics that way. Rote drills, memorization, and force-fed facts were avoided.*

[Isaacson (2007)]

One of the great *tricks of the trade* in mathematical physics is the *thought experiment*, also known as *Gedankenexperiment*. In such an experiment, a mathematical physicist poses a hypothetical question, thinks of some conjectured laws of physics, and works out, either in the mind or on paper, what the answer to the question should be according to those conjectured laws. It's a powerful way of exploring theoretical concepts at no cost in equipment. For instance, mathematical physicists will occasionally discuss General Relativistic spacetimes with closed timelike curves, entertaining the possibility of time travel [Gödel (1949)].

Einstein is known to have used thought experiments during the development of Special Relativity and General Relativity. An exceptionally important example of this involves the *Principle of Equivalence*, that led Einstein to reinterpret gravitation as free-fall in curved spacetime. Einstein imagined himself in a broken lift falling down a shaft [Frank (1947)]. As the lift fell, he would appear to lose weight. In essence, gravity would appear to have disappeared and for the brief moment that the lift was in free-fall, Einstein would feel as if he was a free particle at rest in an inertial frame. This astonishingly simple thought experiment led eventually to General Relativity. Although the thought experiment could be done in seconds, converting it into a proper theory was a hard and tortuous theoretical journey, even for Einstein. Not only was he pressed on all sides by rivals such as Gunnar Nordström, Einstein had to learn sophisticated mathematics to encode his intuition into hard equations [Isaacson (2007)].

Clearly, if such an intuitive approach helped Einstein, it stands to reason that the encouragement of visualization and the development of intuition should be considered seriously , as well as the development of mathematical skills, in schools and Universities. Of course, that is precisely what Pestalozzi believed.

## 11.7 Was Einstein ever wrong?

Although he did great things in mathematical physics, Einstein did make scientific errors and misjudgements. Fortunately, he had the personality to acknowledge some important ones and amend them.

### *The Cosmological Constant*

One critically important concept that lies at the centre of the debate in modern cosmology is the so-called *Cosmological Constant*. When Einstein constructed his theory of General Relativity in 1915, it was generally believed that the Universe was static. When he applied General Relativity to cosmology, Einstein concluded that the theory predicted that the Universe was unstable and would collapse. So in 1917 he introduced a counter-term in his equations of motion involving a really tiny constant, the *Cosmological Constant*, that supplied an effective *outwards* force on large distance scales, effectively counterbalancing the gravitational collapse prediction. Essentially, Einstein stabilized his model.

As it turned out, it was eventually shown by the observational cosmologists Lemaître and Hubble (independently) that the galaxies on the largest observational scales were receding[2]. When he heard this, Einstein became convinced that introducing the *Cosmological Constant* was a mistake and refrained from using it after that. There is evidence that he called the introduction of the *Cosmological Constant* the greatest blunder of his life [O'Raifeartaigh and Mitton (2018)].

All of this was before the discovery of what is known as *dark matter* and *dark energy*. Currently, one of the greatest issues in physics and cosmology is understanding the nature of gravitation on the largest scales. The *Cosmological Constant* plays a central role in research in these issues, because it has been associated with the energy density of empty space. No one knows for sure what's going on. Perhaps *you* will be the person to figure it out. It looks as if Einstein was right to introduce the *Cosmological Constant* but for reasons he could not possibly have known about or imagined at the time.

### *Gravitational waves*

In Chapter 21, *Publish or Perish*, I recount Einstein's confusion about gravitational waves, which at first he believed did not exist in General Relativity, his own theory of gravitation. Fortunately, he realized his error and amended the paper he had submitted in 1936 to a journal. Initially, it had asserted that such waves did *not* exist. Now it said that such waves *might* exist.

Gravitational waves were detected in 2016 [LIGO Scientific Collaboration and Virgo Collaboration (2016)].

### *Oops*

When a collaborator told Einstein that he, the collaborator, was taking care to ensure there were no mistakes in a book they were publishing together, because of Einstein's reputation, Einstein laughed and replied

*You don't need to be so careful about this. There are incorrect papers under my name too.*

[Kennefick (2005)]

## 11.8 No one's perfect

I've mentioned elsewhere in this book that Einstein was human, in that he had his faults. For example, you may be dismayed by his treatment of his first wife Mileva Marić [Isaacson (2007)].

Apart from that kind of personality fault, there was, I believe, a significant aspect of his conditioning that teaches any aspiring mathematical physicist two important lessons:

### *The quest for unification*

James Clerk Maxwell first put the idea into the minds of mathematical physicists that different forces could be described by a single theory. He did it by showing how electricity and magnetism, previously thought of as disjoint phenomena, were aspects of the same phenomenon, *electromagnetism*.

That success motivated and continues to motive mathematical physicists. In particular, Einstein spend the last years of his life in a fruitless effort to combine General Relativity with electromagnetism and other interactions, such as those that hold nuclei together. The remarkable intuition that had led him to create General Relativity and many other great theoretical ideas such as the photon concept, however, failed completely. It was as if the ability to think "laterally" had left him. His previously successful methodology proved in the long run to have limited potential.

I have a suspicion why he failed. When you think of it, both gravitation and electromagnetism exhibit macroscopic scale behaviour. We can legitimately imagine ourselves falling with a broken lift down a lift shaft and feeling weightless. That sensation can be experienced momentarily by anyone jumping down from a tree. With electromagnetism, we interact with light also on macroscopic scales. Thought experiments about inertial frames of reference measuring the speed of light are reasonable also. When it comes to nuclear (strong) and weak interaction forces, however, we simply do not experience them in the wide world. Therefore, I would be surprised if any visualization involving those forces had more than marginal value. Does it really make sense imagining yourself falling into a nucleus?

### *Issues with quantum mechanics*

Einstein had a great deal to do with the development of quantum mechanics, right from its earliest days. He suggested the photon-quantum particle concept, interpreted the squared modulus of an optical wave amplitude as a probability density of detecting photons (an idea that influenced Max Born in his probability interpretation of Schrödinger wave mechanics), and influenced Luis de Broglie in his suggestion that material particles such as electrons would be accompanied by waves. Einstein also helped develop *Bose-Einstein statistics*, necessary for the correct quantum mechanical description of systems composed of identical particles known as *bosons*.

Given all that, Einstein ended up sceptical that quantum mechanics was a 'complete theory'. The best known illustration of his doubts was the publication of the now famous *Einstein-Podolsky-Rosen* paper of 1935, usually referred to as *EPR*, in which the authors attempted to show that quantum mechanics did not model physical reality completely [Einstein *et al.* (1935)].

As with his obstinate and ultimately unsuccessful belief in unified field theory, Einstein clung on to a realist belief that underneath the radar of observation (which requires quantum mechanics to explain its data), there is a hidden world based on some form of classical reality. Subtle experiments involving the so-called *Bell inequalities* have made the realist position taken by Einstein and others (the *Hidden Variables* adherents) virtually untenable at this time.

## 11.9 Lessons

I believe that Einstein's last years provide several important lessons for mathematical physicists:

(1) Never stop examining your mind for signs that you have fallen into a conditioned rut. This is particularly important the older you get.

(2) Apply *Nullius in Verba* ruthlessly to yourself as much as to others.

(3) Know when to move on and acquire new conditioned beliefs based on the most recent empirical evidence. Keep up to date with developments.

(4) Realise that past successes do not necessarily give any pointers to the solution of new challenges.

None of this is a criticism of Einstein. Personally, I would be satisfied to have written just one of his many great papers.

# Chapter 12

# Motivate Yourself

*For metaphysic, we have assigned unto it the inquiry of formal and final causes; which assignation, as to the former of them, may seem to be nugatory and void, because of the received and inveterate opinion, that the inquisition of man is not competent to find out essential forms or true differences; of which opinion we will take this hold, that the invention of forms is of all other parts of knowledge the worthiest to be sought, if it be possible to be found.* ***As for the possibility, they are ill discoverers that think there is no land, when they can see nothing but sea.***

Francis Bacon [Bacon (1893)]

In the next theme, *Practicalities*, after this chapter, I go over just that: the actual becoming, and practice of being, a mathematical physicist. Those chapters are based on the assumption that you have been enticed by my sales pitch for the subject and are considering finding out more about the discipline.

Being a realist in this respect, and knowing that conditioning is hard to alter, I cannot rule out the likelihood that you are not sure at this point of the book. It's all very well me advising motivation, but you may not know how to go about acquiring that motivation. In this chapter I give you some guidance in that critical respect. As in other chapters, I'll modularize my thoughts, meaning I'll write out a string of observations and comments, in no logical order. Perhaps taken as a whole, you will have some encouragement to go further into the subject of mathematical physics.

## 12.1 It can be for you

Just because you may not have heard about mathematical physics before is no reason to think you could never become a mathematical physicist. The problem is that schools don't teach mathematical physics as far as I know, you don't come across the discipline in the news media, and you don't meet mathematical physicists every day. It's not a career path normally advised in schools, colleges, or even Universities.

It's true that there have been some entertaining films, documentaries, and biographies about Einstein, Hawking, Oppenheimer, and other mathematical physicists. However, the focus in every case was the individual, not the subject of mathematical physics *per se*.

I don't think any mathematical physicists ever decided to go into the discipline at an early age, because they would have had no idea that such a career existed. When I was a child, most of my friends wanted to be train drivers, cowboys, or ice-cream sellers (on the assumption that then they would get their ice-creams for free). When I was six or seven, I wanted to be a pilot (flying a Hawker Hunter jet fighter specifically), because my uncle had been in the Air Force during the Second World War and I was fascinated by his memorabilia. I think mathematical physicists gravitate into the subject through a series of unrelated experiences and decisions that separately don't point towards the discipline, but collectively end up pushing the individual right into it.

You and your friends will probably share many interests that have nothing to do with mathematical physics either, such as computer gaming, music, sport, socializing , and so on. It's possible that you're worried that if you mentioned to them that you were thinking about becoming a mathematical physicist, they might laugh at that ambition, or tell you that it's unrealistic.

Don't listen to such negative views. What you will almost certainly discover is that the great friends that did everything with you will sooner or later each find their own niche in life and perhaps go their separate ways, never to be seen again. Will you think any less of them if any of them tells you that they want to become an airline pilot, a police officer, or even a writer? You would not be much of a friend if you laughed at that. Likewise, good friends would surely admire you for having any long-term ambition.

When it comes to your own life, don't let others influence you in your personal choices.

## 12.2 Watch science documentaries

These days, an important factor in making science attractive to the general public is the quality of science documentaries. Many years ago, science documentaries on television would be little more than glorified lectures done in black and white on low resolution displays. Nowadays they can be greatly entertaining dramas, with grand music and glorious computer graphical effects on high definition flat screens. Moreover, there has been a veritable flood of wonderfully stimulating discoveries in virtually every field of science, ranging from Astrophysics and cosmology to the neurosciences (how brains work). Mathematical physicists can contribute in all those areas.

## 12.3 Subscribe to science magazines

A good way to develop interest in science is to read popular science magazines such as *New Scientist*, *Scientific American*, and any of several popular Astronomy magazines.

Mathematical physics is too specialist to have its own popular journals. Radical new theories that occasionally crop up in mathematical physics will usually be mentioned in the popular science magazines I've just mentioned. For example, *Modified Newtonian Dynamics* theory, referred to as *MOND*, is a recent attempt to explain dark matter by modifying the Newton-Cavendish inverse square law of gravitation, rather than Einstein's General Relativity. Popular science journals have quite properly given this alternative theory fair coverage.

From this, you can be confident that mathematical physicists are involved in one way or another with modern Astronomy and Physics. Finding out about recent empirical discoveries and alternative theories by reading popular science magazines is recommended.

## 12.4 Read science biographies

I discovered when I was at school a biography on Niels Bohr, the mathematical physicist who developed the planetary model of atoms. Reading that book was the first time that I realized that there were people who did nothing except *think* about physics for a living. There were quite a few amusing anecdotes about Bohr in that biography. He came across as a relatively normal person with an unusual career.

If you want a particularly entertaining autobiography about a mathematical physicist, you could do no better than to read Richard Feynman's book *Surely You're Joking, Mr Feynman* [Feynman (1985)].

## 12.5 Start scribbling

I imagine you have been given some maths problems in your time. I imagine you believe that solving them should be done in the classroom.

As I see it, that may well be the worse place for solving problems. Get a pad of paper, a good pencil, and start solving maths and physics problems as often as you can, at home, on the bus, on the train, and, yes, at school if you have to. But definitely, don't "switch off" your mind from your problems outside school.

I think a crucially important point here is to *solve your problems on your own.* Don't ask Jimmy or Annie how they got the answer. Go away somewhere where no one can bother you and have a go on your own. It doesn't matter if you fail. That's not the issue. If you can't solve a problem after you've had a really really long go by yourself, then *that's* when you can go seek advice from others. But do make sure you have had a go and are not quitting too soon.

## 12.6 Walk and think

Solving problems is not done just on paper. There's no point in having a pad of paper and a pencil if you have no ideas. Ideas have to be encouraged; they have to be given the right conditions before they leap out of the depths of your mind.

There is now evidence for the claim that sleep (of the right kind) can promote creativity [Walker (2017)]. When people sleep, they are not conscious in the way that they are when they are fully awake. In particular, the stage of sleep known as *REM* (rapid eye movement) sleep has been identified as the one where free-association of memories occurs, leading to creative insights and dreaming.

In an analogous way, recent research has shown that *walking* also has a beneficial effect on creativity [Oppezzo and Schwarz (2014)]. According to the authors of that report:

"*Walking outside produced the most novel and highest quality analogies. The effects of outdoor stimulation and walking were separable. Walking opens up the free flow of ideas, and it is a simple and robust solution to the goals of increasing creativity and increasing physical activity.*"

There are historical precedents for the idea that walking assists creativity. Charles Darwin the evolutionary scientist, William Wordsworth the poet, and Freidrich Nietzsche the philosopher, all used walking to stimulate their minds [Young (2015)].

I would recommend regular walking *on your own* in quiet places as a low-cost, immediate access activity that will give you the opportunity to relax, tone up your legs, and allow your mind to meander over any maths and physics problems that you have been studying. One hour's hard study at a desk followed by a good walk, and then a return to the desk, is a great way to study and do research.

A major advantages of brisk walking is that you do not need to carry any equipment: no paper no pen, no tablet. Try to use your mind alone to visualize problems and their solutions. If your experience is anything like mine, you will find that you'll get better and better at thinking out potential solutions to your problems during your walks. They don't all work, of course, but occasionally, they do.

## 12.7 Join scientific clubs

Real scientists don't hide away usually. Sometimes they do, but meeting other scientists regularly is a good way to avoid several hazards. One hazard is that isolation can make you out of touch with recent developments.

Another hazard is that the problem you have just solved after three month's isolation and intense work was solved two months ago and widely discussed in the science club you didn't attend during those three months.

The moral is to keep in touch with other mathematical physicists if you can.

## 12.8 Attend Open Days

Universities and colleges these days are all eager to recruit students. Open Days are now held throughout the academic year, giving school pupils the opportunity to visit departments and talk to staff. I would recommend attending Open Days in mathematics departments, where you may get to meet actual mathematical physicists and find out what they're currently working on.

The only real problem here is the cost of travel to far-off departments. This may mean that you have to do some research online in order to choose the right departments for you to visit. For instance, if you've been stimulated by science documentaries on black holes, it's not going to be much use to you if you travel a long way to a department that researches only in non-relativistic quantum mechanics.

## 12.9 Invite science speakers

School Careers Guidance teachers and School Society organizers should be pro-active in inviting mathematical physicists to talk about their work. Most academics would be pleased to get an invite, for two reasons. First, giving a talk to schools should not involve research level technical details, such as mathematical equations. An overall, general discussion is all that will be required. Any academic worth their pay should be able to do that off the top of their head with minimal preparation and do it well. Therefore, I don't think *pressure of work* stands up as a reason for declining to give such a talk. The only problem as I see it is *timing*, because University lecturing times generally overlap school times.

A second reason is that *it looks good!* Any academic who informs their department that they have given a talk to a school will be applauded for their positive representation of the University and their department.

As far as financing travel and hospitality are concerned, giving talks to schools is arguably a social responsibility of Universities and Colleges, so there is a chance that travel and subsistence costs can be met from their departmental budgets, rather from the schools being visited.

## 12.10 Communicate online with scientists

There are numerous online fora such as *ResearchGate* that allow participants to ask "scientific" questions, such as "Can gravity be repulsive?", or "Does the speed of light change over time?" There will generally be a flood of answers from all sorts of individuals. Some will be experts in their field whilst others will display an obvious ignorance.

My experience has been positive and negative. On the positive side, I have read some fine and scientifically consistent replies to good questions. On the negative side, there have been some replies (not to me, I should add) that have demonstrated unwarranted abuse and hostility towards the questioners, rather than addressing the questions they asked. That is an unfortunate aspect of online dialogue that is prevalent in social media and which leads me now to the conclusion that such media may best be avoided.

## 12.11 Avoid pseudoscientists and charlatans

One of the biggest non-technical problems facing you as a beginner in mathematical physics will be to identify reputable scientists and those that are, quite frankly, best to avoid. It is all too easy these days to strike up an email correspondence with people you have never met. Some of them will be scientifically unsound. As with friends, if you fall in with the wrong crowd, you will have problems sooner or later. I should give you some advice about spotting people to consider avoiding when it comes to mathematical physics. I regret to say this but out there in internet land are numerous pseudoscientists waiting to convince anyone who clicks on their websites that they have the secrets to the Universe and no one else has.

How can you spot scientific frauds and charlatans? Some of them may indeed have good intentions and are not deliberately attempting to deceive others, but they have perhaps fallen into a conditioned mental rut where they think they are right and everyone else, particularly the *Establishment*, is evil, wrong, and conspiring to get them.

My best advice here is to take *Nullius in Verba* literally. Look carefully at what people write online. I will refer to people to avoid as *Problematicals*. Here are a few clues for spotting *Problematicals* online:

**1.** *Problematicals* often have websites with strange, mystical logos all over the page. These are designed to induce a sense of wonder and awe in the unwary visitor to the website.

**2.** *Problematicals* invariably do not adhere to *Nullius in Verba*, so their websites make unverified assertions with no reference to supporting data;

**3.** *Problematicals* often make derogatory comments about established proper scientists, reputable organizations such as CERN (that runs the *Large Hadron Collider* experiments), long established reputable journals with proper refereeing standards, and well-documented and successful theories such as quantum mechanics and General Relativity.

**4.** *Problematicals* often refer to conspiracies by the scientific establishment to suppress the *Problematicals*' theories.

**5.** *Problematicals* often self-publish strings of papers that purport to explain so many things that regular science cannot.

**6.** *Problematicals* don't publish in properly refereed, reputable journals.

**7.** *Problematicals* sometimes publish in predatory journals. These make money for the publishers by promising to publish virtually any submitted paper, at a cost of course.

**8.** *Problematicals* often give unusual, idiosyncratic names to concepts that they invented, sometimes even their own name.

**9.** A surprising percentage of *Problematicals* seem to have developed a "General Theory" of this or that, motivated probably by Einstein's history. He first developed the Special Theory of Relativity, and then the more comprehensive General Theory of Relativity.

**10** This is perhaps the most unfortunate fact of all. If you do look carefully into what *Problematicals* have theorized, you invariably find that you can do nothing with their ideas. Their "theories" tend to be a mass of unprovable assertions drawn from thin air to "explain" whatever takes the

*Problematical*'s fancy. Science is not stupid. If a *Problematical* did come up with a theory that actually gave new, verified predictions that other theories could not, reputable scientists would be the first to acknowledge that (or they should be).

Science is not conspiratorial in nature. There may be pockets of conditioned blind adherence to some enterprises such as *String Theory* and *Many Worlds*, but there is a long-term fundamental honesty in the scientific community. *Nullius in Verba* sees to that. Sooner or later, frauds and charlatans are found out, although in the short term, it's possible to be taken in. To make sure you are not deceived, do your homework: check the credentials of anyone you are communicating with online. A good way is to look where they have published their theories.

## 12.12 Never give up, never surrender

The motto "*Never give up, never surrender*", from the film *Galaxy Quest*, is worth remembering when times are hard and you feel you are losing. Sometimes, persistence and hard work pay off, and a bad situation can be turned around. One thing is for certain, however: if you don't try, you won't achieve.

This applies to study as much as to any aspect of life. If you want to become a mathematical physicist, you may well face tough times, perhaps failing exams or running out of time and money at critical points. It's at such places in your life that your character will be tested. Do you have what it takes to stick it out? Sometimes, miracles do happen, as this next anecdote illustrates.

**Determined**

In my time as a personal tutor to numerous mathematical physics students over the years, I encountered apparently weak students who were strongly fired up with enthusiasm for their subjects but who started their degrees badly. Many of those students got better and better at those subjects and in some cases, outperformed in the final exams students who had started their degrees brilliantly but then lost interest and motivation by the end, a few years later.

The structure of our degree course in mathematical physics involved an initial year that was for acclimatization purposes and didn't count towards the final degree classification. There were exams in the second year and then in the third year. Those exams counted.

Now put yourself in the position of a second year student who had done miserably in their second year exams. What motivation would they have to do well in their third year? Some such students actually got worse and ended up with minimal degree classifications.

At one point, I became so concerned that I analyzed four years' worth of second and third year exam performance in mathematics and mathematical physics, involving several hundred students. The results were encouraging and suggested (to me) that *nothing is inevitable: it's down to the individual.*

To explain what I found, you have to understand the UK classification system. In those days, final degree award was in terms of *class*, decided by your actual percentage marks overall:

| | |
|---|---|
| 70 – 100% | gave you a First Class degree, |
| 60 – 69% | gave you a Two-One degree, |
| 50 – 59% | gave you a Two-Two degree, |
| 40 – 49% | gave you a Third Class degree. |

I found that of the people who got a First in their second year, about half of them got a First overall and the rest went down, usually to a Two-One. Of those who got a Two-One in their second year, about half ended up with a First and some went down to a Two-Two. Of those who got a Two-Two in their second year, about a half went up up, mostly to a Two-One. A very small number got a First.

Here's the miracle. There was one exceptional girl who got a Third Class in her second year. She was so determined that she worked and worked away, motivated by her interest in the subject. She came out with a First Class Honours degree in mathematics a year later. That was a sensational example of long term motivation beating a poor start.

I was able to use my analysis to show my students that *nothing is certain.* If you are motivated, you should work away regardless of what the situation looks like right now.

# Chapter 13

# Study, Sleep, and Exams

*To Sleep, or Not to Sleep, That Is the Question*

[Ayas *et al.* (2013)]

In this chapter, I discuss studying and the crucial importance of getting a good night's sleep. If you're a student, it won't be much use to you if you get that sleep *during* an exam. If you're a mathematical physicist, a good night's sleep will help concentration and creativity.

## 13.1 Study and environment

In my experience, thinking properly requires the right environment. If you're surrounded by loud neighbours, or are constantly picking up your mobile phone to see the latest email, text, or *Whatsapp* message, don't expect to come up with any great theory in a hurry, or to study efficiently. Find a quiet seat in a Public Library and throw your mobile phone in the river on your way there. Avoid your friends for a couple of hours and certainly don't start a conversation with that attractive person on the other side of the table. If you do, say goodbye to your thinking time and to your day's work.

It's not that I'm antisocial, but I discovered that socializing is simply not helpful as far as thinking goes. That conditioned me to an interesting way of studying during my undergraduate and postgraduate years. Instead of sitting in my assigned room, I would wander off around whatever University campus I happened to be in, looking for unusual and remote corners to study in. I would go to libraries I had never been to before, such as the Philosophy Department's library, and do some work there. The environment in such places was generally conducive to thinking and to doing calculations. No one knew me there and people just ignored me.

### Time and Place

I don't know about you but I am susceptible to the *atmosphere* of a place or building. I remember standing below the *Occulus*, the circular opening at the top of the great dome of the Pantheon in Rome and being overcome with a sense of awe. I felt as if I was seeing into the mind of the person who had designed this marvel of construction. I imagine it's the same with the Pyramids, the Taj Mahal, and the Acropolis. Edinburgh is like that. I spent four wonderful years there.

One dark and cold night as a final year student, I found myself sitting alone in a common room, high up in *Mylne's Court*, studying hard on some mathematical physics problem sheet.

Mylne's Court is an impressive old building, renovated to a high modern standard but keeping its essential historical character. When I was there, it had recently opened as a University student hall of residence. It occupies a prime position on the great hill of the City known as the *Mound*, right on top of which is the famous castle. Mylne's Court is an example of a seventeenth century skyscraper, soaring high into the air, with about ten levels and many windows. I was most fortune to have lived as a student in such a place.

Mylne's Court had an air about it that resonated with my ambition regarding mathematical physics. I found working away in various odd corners of that old building most conducive to my studies. It was there that I became conditioned for life to getting up before dawn and doing a couple of hours' work before most people were awake and before breakfast.

On this occasion, however, I must have been worried about the forthcoming summer exams, still three or more months away, so I found myself having a late night study session.

It was going well. It was already quite dark that cold, damp, winter night. It was also very late into the night and a strange, chilly silence hung all around. There was no noise from the distant streets. The whole Universe apart from me seemed to be fast asleep. I had gone to one of the communal kitchens, where I could make coffee and work undisturbed at the table there. That room was actually over a *pend*, one of those narrow staircases that characterize the architectural style of the old buildings here, running all the way from the Castle down towards Princes Street, the great main street of the city.

I looked up from the table where I sat, with all my papers scattered around, for some motion had caught my attention at that moment. Although it was late and cold, I had previously opened wide a small window that overlooked the pend below that ran towards Princes Street. Doing mathematical physics requires a clear head, after all. Through that window, I could see a proper Edinburgh dense fog, masking the distant lights of Princes Street. What had caught my attention was the fog rolling over the window sill, seeping eerily into the room.

Suddenly, I felt a strange chill running down my spine. This was a really old building, at the heart of an ancient town. Famous people, long dead, had walked the streets here. The great David *Hume*, one of only three philosophers I rate highly (the others are Socrates and Confucius) lived very close by for a time in 1769. I felt at that moment as if I had stepped back in time hundreds of years.

But that was not all. Right at that moment, someone, somewhere, started to play the *bagpipes*.

Now, I don't know about you, but the bagpipes are what can only be described as a *binary* instrument. You either love them or loath them. When they are played properly, I love them. They are superb in creating that particularly Scottish atmosphere that encapsulates perfectly the essence of their culture. That person playing the bagpipes somewhere in that huge building knew how to play.

At least, I think it was a person playing. But late at night? In a built up area?

No. I'm sure. It couldn't have been a ghost. Could it?

As a trained scientist, I don't believe in ghosts. But my imagination does.

I left that room in a hurry.

## 13.2 Consequences of sleep deprivation

Every day, the time comes for you to wake up. How your day works out may well depend on how well you've slept the night before. This will be particularly true if you're someone who is going to rely on their thinking processes during the day. If you're going to be a mathematical physicist, you should regard a good night's sleep as a number one priority. This goes for any sort of creative activity involving your mind.

If you look at it superficially, sleep looks like a useless activity. Let's say you sleep seven hours a day. That means that by the age of twenty four, you will have slept *seven years* in total. It's probably more than that because teenagers need between eight and ten hours' sleep per night according to sleep researchers. Why do we need to sleep?

This is a simple question with no simple answer. One line of research suggests that it is a clearance mechanism for the brain, which I find a helpful idea (it gives me a motive to go to bed when I don't want to). Let me give you an analogy. I have a PC (personal computer). I built it myself. It's fast, has lots of RAM, huge hard disk space, and three monitors. I use it extensively for writing my papers and books. In some respects, my PC is like a brain. It can do several tasks concurrently, but it's not perfect. Every so often, it starts to get clogged up. Whenever I install or uninstall an app, some junk gets left behind. Whenever I look on the internet, some unwanted code gets through, in the form of cookies and trackers. Eventually, my PC starts to underperform.

My solution is to do regular maintenance once a week to clear the junk. Every Sunday morning, I apply a well-known free app that clears my PC's *registry* of extraneous links. Some weeks I've found several hundred broken links that should be removed.

There is good evidence for the suggestion that one of the key functions of sleep is analogous to the clearing of a computer's registry, but applied to the brain. Recent research on mice shows that during sleep, the spaces between brain cells increase by more than 60%, allowing toxic waste products to be removed efficiently from the brain [Xie *et al.* (2013)]. The researchers of that article conclude that

"*...the restorative function of sleep may be due to the switching of the brain into a functional state that facilitates the clearance of degradation products of neural activity that accumulate during wakefulness.*

Provided you have had enough proper sleep, you should be well prepared for high level mental activity during the day. However, if you are long term sleep deprived, all sorts of medical problems can arise, such as weight gain and a weakened immune system, in addition to adverse effects on your thinking processes. Here's a list of the potential adverse consequences of sleep deprivation in teenagers [Garey (2019)]:

(1) Increased risk of injuries, such as "fall-asleep" car crashes.

(2) Lack of self-control over emotions, impulses, and mood.

(3) Increased aggression, impulsivity, and short-temper.

(4) Increased use of stimulants containing caffeine and nicotine.

(5) increase of self-medication with alcohol.

(6) Impaired judgement and poor decision-making.

Over the years, I've observed the chronic problems caused by sleep deprivation in myself and others. Usually, when I've not had enough sleep, I feel jaded, burnt-out, and lacking in initiative. Under those circumstances, doing anything more than routine work becomes near impossible. This is consistent with a recent theory linking sleep and creativity [Yong (2018)]. If that theory is borne out, then it will be clear why good sleep is important to a mathematical physicist.

## 13.3 Sleep mythology

There's a lot of mythology, or fake news if you like, about sleep. Here are some common beliefs discounted by recent research [Robbins *et al.* (2019)]:

(1) Your brain and body can learn to function just as well with less sleep.

(2) Lying in bed with your eyes closed is almost as good as sleeping.

(3) Exercise within 4 hours of bedtime will disturb sleep.

(4) Adults sleep more as they get older.

(5) Drinking alcohol before bed can improve your sleep.

(6) Snoring is harmless.

On those rare occasions when I've worked well into the night, and once or twice, completely through to the next day, the short term gain of doing some work in those night hours was always cancelled by being unable to do any meaningful work afterwards. On one occasion, when I had to reach a publisher's deadline and decided to work throughout the night, I ended up stalled, like a car engine driven badly, and could do no more work for several days. My sleeping pattern was destroyed and I felt like a zombie (awake but not meaningfully conscious). The result was that I had to email the publishers for an extra two weeks' extension (which they gave without any problem).

## The Hour of the Wolf

I didn't know it as the *Hour of the Wolf* at the time, but when many years later I saw Ingmar Bergman's film with that title, I knew I had experienced what he was driving at. According to that film, the *Hour of the Wolf* is "*The hour between night and dawn. The hour when most people die, when sleep is deepest, when nightmares are most real. It is the hour when the sleepless are haunted by their deepest fears, when ghost and demons are most powerful, ...*".

I'm referring to the last few nights before I completed my doctoral thesis. I had worked for close to three years on the quark-parton model of strong interactions, and my self-imposed deadline for submission was looming. Long sessions during the day were the rule, but it was beginning to look problematical as to whether I would complete in time. So I started to work throughout the night.

I've mentioned before that location is one of those ingredients that can make or break creative work. A tedious environment dulls my senses, whereas a bit of excitement I find stimulating. I've often found new thoughts flooding into my mind when on a train, for example (but it never helps if you have people to talk to, as that's distracting).

I decided to work in the departmental library throughout several nights. In those days, every research student had a key for twenty-four hour access, and there were no restrictions on how long you could stay. Moreover, by about ten o'clock at night, the large, well-stocked library would be quite empty, apart from me. I would sit at one of the huge tables with all my papers and books scattered around on it, with a solitary light focused on me and my work. Outside that narrow circle of light, the shelves and walls of the library would fade away into utter darkness.

Under such circumstances, one of two things would happen. Either I would find myself concentrating deeply on my work, or else the reality of the situation would wash over me and I would succumb to *the Hour of the Wolf*. How can I describe that last phenomenon? It's a stage of mental activity where your consciousness switches from whatever you're concentrating on (which we can think of as the *inner* focus) to real and imagined external events (which we can think of as the *outer* focus). Essentially, a person gets distracted easily and work becomes pointless. Any slight noise, such as a clock ticking, or a tap dripping from a nearby washroom, sounds like an avalanche.

In those days, security was relatively minimal, but it was there. I soon found out that although the building appeared deserted at night, a security guard would be on patrol somewhere. One of the most disturbing aspects of the Hour of the Wolf was the sudden, noisy appearance at three o'clock in the middle of the night of a startled security guard, unlocking the door to what he thought was a deserted library, only to see an unshaven, dishevelled research student looking up in sleepy confusion from the side of an enormous table over which were scattered the messy pages of unfinished calculations and a half-written thesis.

I thought I had learnt my lesson then and had resolved never to experience another *Hour of the Wolf* again in my life. But publisher's deadlines have to be met. There's every chance I'll be editing this paragraph in the middle of the night in about three week's time.

My best advice: avoid your own *Hour of the Wolf* at all costs.

## 13.4 Sound sleep

Here are five reasons for getting a good night's sleep, especially before examinations [Walker (2017)].

(1) sleep enhances memory and refreshes learning ability. You need sleep *after* study in order to store properly the memories involves. You also need sleep *before* study in order to prepare your brain for learning.

(2) research shows that proper sleep can enhance creative ability significantly.

(3) short sleep appears to have significant connections with major diseases such as Alzheimer's disease and cancer.

(4) short sleep reduces life expectancy and quality of life.

(5) short sleep amongst workers is economically bad for companies and nations.

My best advice to any student is to set up a sound sleeping regime as a matter of priority, *before* embarking on any revision for examinations. The same advice applies to doing research. In this context, I discovered that there is a lot of mythology about Einstein's sleeping habits, claiming that he slept ten or more hours a night. There is no reliable documentation for that belief. According to one biography, Einstein slept eight to nine hours regularly, and believed that being well rested and healthy was instrumental to his productivity [Isaacson (2007)].

## 13.5 Exams

Exams are critical events in a student's life and require extensive preparation for. One of the hazards of the game is the urge to do late night study the night before an exam. Here are some anecdotes involving sleep and exams.

**My Last Exam**

There I was, the night before my very last mathematical physics Finals exam, sitting at my desk in my Mylne's Court Hall of Residence bedroom, worn out and ready for sleep. After three year's hard work doing mathematical physics, I was poised to end it all at noon the next day, having finished a three hour exam on Einstein's great theory of General Relativity that had started at nine o'clock in the morning.

It had been a stressful four weeks of exams, in the middle of a hot summer. Half way, I caught a touch of sun-stroke that had knocked me out for a couple of days. That was my fault completely. Given a couple of exam-free days right in the middle of those four weeks, I had done exam revision in the park known as the Meadows, sitting out in the hot noon sun with my books and lecture notes spread out on the grass, with no sun hat or shade. That was sheer madness. If you want to experience hell on earth (pardon the expression), do what I did and revise in those conditions.

Back to my story. As I said, it was nine o'clock in the evening before my last exam. I had spent the entire day revising, revising, revising, and by nine I was ready to quit. You will know what that's like. We're all like batteries really. An AA battery will run at one and a half volts for a reasonable time but will eventually run out of power. So will you if you overdo anything.

It was not dark yet but already I could feel the sleepiness start to creep into my brain. A nine o'clock exam the next morning meant a good night's sleep was in order.

But ... something nagged at me. Perhaps at that critical moment, my subconscious came to the rescue and stopped me from quitting my desk. As the nearby church clocks struck nine, I spontaneously decided to stay and review two past paper exam questions, set some years previously, in General Relativity.

I opened my folder and went over those two questions. I had written out complete solutions to them weeks before, so this time it was just a matter of refreshing the details, the short cuts, the technical moves, the tricks of the trade. If you tackle a mathematical physics problem, you will find that it's like a boxing or fencing match against a devious opponent. The problem is your opponent and to defeat it, you have to outwit it with all the fancy mental footwork that you can muster.

Ten minutes later, that was it. I had had enough for sure. What those ten minutes gave me was not so much knowing those two problem solutions inside and outside, forwards and backwards, but having the reassurance that I had done everything I could for the exam, and even gone that extra mile. So I went to bed relatively calm.

The dreaded moment came. There I was, just before nine the next morning, sitting at a desk in the middle of rows and rows of other desks, having filled in all my details on the answer book cover, such as name, rank, and serial number (I never got any extra time for having to write out my long surname).

Then came the dreaded words: "*The exam starts now. You may open your examination question book and begin.*"

I did that and started to read the questions. That's important. Don't start answering questions until you have a strategy. Read all the questions first, then decide on the order in which you should answer them, assuming you have a choice.

In this particular exam, we had to answer any four questions out of six possibilities. So I read the six questions.

**Two of them were precisely the two questions I had last-minute revised twelve hours before.**

I don't need to tell you which four questions I decided to answer. You will guess, I'm sure, that two of them were those last-minute revisions that I knew inside-out and back-to-front.

There are two points about this story.

First, there must have been a touch of laziness in the examiner, who had decided to use two old exam questions instead of inventing new ones. That was my good luck.

Second, there have been moments in my subsequent career when I have reflected on this event. Suppose I had not revised those questions. Then perhaps I might have done badly in the exam, perhaps even failing it. Perhaps then I would not have obtained a good Mathematical Physics degree, and so not gone on to do a doctorate, and so then I would have had an entirely different sort of life. Should I feel guilty about having it so easy in that final exam?

Part of me says *yes*, until, that is, I realise the essential truth of the matter: *I* made my own good luck by choosing to stay at my desk the previous night, when common sense dictated that I should have quit and gone to sleep. Whenever I remember that, my conscience becomes clear.

Humans are sometimes pretty strange creatures. What appears reasonable to some may appear unreasonable to others. Here's a case in point. Make up your own mind as to whether you think the complaining students had a point.

## It's Not Fair

For years, people in the maths department where I ended up working for several decades set their exams more or less how *they* felt. Some colleagues devised new questions every time. Others would throw in, occasionally, a couple of questions from past years. As for me, I found that setting good exam questions was a real chore. You have to get the balance right, making it relatively easy for a weak student to get some points at least, and giving the geniuses a chance to get one hundred percent (it can happen). I would occasionally reuse a good question, but I would generally alter it a bit.

On this particular occasion, there was a complaint from several students after the results came out for an exam given by another colleague of mine. The nub of their complaint was this. They said that: *Some of our fellow students went over last year's exam and its solutions, but we did not. This year, two of the six questions on this year's exam were from last year. Therefore, we who did* not *go over last year's exam were disadvantaged.*

Can you believe it? The complaint was that *students who did not study found themselves at a disadvantage over students who did.*

We live in strange times. Years ago, such a complaint would have been thrown out as idiotic. What did my department do? Our Exam Committee issued an edict, that in future, no colleague was to recycle an exam question from the previous year's exam.

I suppose it was just my good luck that such an edict had not been in force when I sat my last exam (see previous anecdote) all those years ago.

An important point to think about, regarding these anecdotes, is to never underestimate the role of *Black Swan* events in your life. A Black Swan event is something that comes as a surprise and has unexpected and potentially catastrophic consequences [Taleb (2010)].

My next anecdote is about a good student of mine who was in the middle of his second-year mathematics exams. Those exams counted towards his final degree classification (the first year exams never counted in that sense, being a "warm-up" for the real thing later). Let's call that student *Worried Lad.*

## Overdoing It

Sitting at a desk seems to sum up my life as a mathematical physicist. There I was, one bright day, in my office, at noon, at my desk, and eating my lunch. Lectures had been over for a couple of weeks and students were busy with their exams. So peace and quiet were the order of the day.

Until, that is, *Worried Lad* burst through my door. Did I say *burst through*? He almost tore the door off its hinges.

He came in, with a dreadful look of despair on his face, and slumped down in a chair next to my desk.

Right away I knew something was up. "*He's in the middle of his second-year exams*", I thought, "*So what's up*"?

As his personal tutor, it was my business to sort out his academic problems. I always advised new personal tutees about that. I told them I couldn't advise them about other issues, such as lending them money or sorting out their social life, but anything to do with academic matters was fair game. So I asked him what the problem was.

He blurted out that he had just finished an exam that morning and it had gone terribly badly. He had prepared for it, but to no avail. He just couldn't understand what had gone wrong. Worse, there was another two-hour exam due to start at two o'clock, and he was in total despair now.

I looked at my watch. It was now just after twelve. I would find out what had gone wrong and try to get his mind back on track for the two o'clock exam. It was just a question of morale. *Worried Lad* was a good student so I needed only to restore his confidence and that should do the trick.

So for the next half hour, we discussed what had gone wrong that morning. Did he go over past papers? *Yes!* Had he missed any lectures? *No!* When did he start revising? *Weeks ago!* And so on.

Half an hour later, we had come to no conclusion. Everything seemed to show that his morning exam should have gone excellently. As I said, *Worried Lad* was a good student.

And then he said it.

"*I don't understand it. Everything was perfectly clear to me when I finished revising at* ***two o'clock this morning*** *and went to bed.*"

The explanation for his disaster that morning came like a flash of lightning. Inadequate sleep had ruined that morning's exam for him, and he was quite unaware of that reason. It had taken a whole half hour's discussion for him to give that vital information. Essentially, he had gone to bed after two that morning, woken up at seven, in order to start an exam at nine. So he had less than five hours in bed, which probably meant that he had had less than five hours sleep. And he could not see it.

I immediately took the following action and advised as follows. Stop this discussion! Calm down! Go to the two o'clock exam and do your best. Afterwards, go back to your Hall of Residence, get a meal, and go to bed early, so as to get seven or more hours good sleep. Never work so late again in the run up to exams.

There is an ending to this story that speaks volumes about the importance of sleep. Despite his bad exam experience that morning, *Worried Lad* did not fail his overall exams and went on into his third year. At the beginning of that third year, we had the standard personal tutor-tutee meeting, where we discussed the past year and planned his final year. I made sure he kept in mind the necessity of a good sleep regime. He reassured me that he had learnt his lesson.

But, sad to say, the same thing happened in the May of his final year. Once again, he came to me in desperation after a bad morning exam, with much the same story as the year before. Somehow, he had not absorbed the lesson that the previous year should have taught him. Yes, *Worried Lad* got his degree, but along the way, his patterns of sleep during revision played havoc with his exams.

## 13.6 Larks and Owls

By now, you will be aware that I go on a lot about conditioning, meaning, the way that our thinking becomes dominated by acquired habits. However, that is not the whole story. Research by neuroscientists has shown that there is a strong *genetic* component in our sleeping habits. It looks as if some people are naturally *owls* and others are *larks*. *Owls* are people who prefer to work late into the night and get up late in the morning, whilst *larks* go to bed early and wake up early. I discovered at University that

I was a lark. It's possible *Worried Lad* was a lark who forced owl-like behaviour on himself and suffered in consequence.

There is an important issue here that teachers should think about. If exams are set too early in the morning, they might affect owls more than larks. Moreover, there is research by neuroscientists that supports the notion that teenagers particularly have a predisposition to sleep later into the morning than adults. I won't give references here to such research, as it can be found readily online. The essential point here is that the evidence suggests that teenagers who sleep late are not "lazy", but behaving according to the way their brains are operating at that time of life. A solution would be to start exams later in the day, or even, possibly, abolish exams altogether.

My personal view is that giving a set time for an exam has no value, unless you think people should behave like robots doing work as fast as possible. Real life is usually not like that. Certainly, you pay a plumber to fix your bathroom leak as quickly as possible, because it's inconvenient to have no water. But on the other hand, if you are having an operation on a brain tumour, you would be foolish to demand speed from your surgeon so as to reduce her labour charges. You would want the operation done *properly*, in as long a time as the surgeon requires.

The same goes for mathematical physics. Given the great problems of the universe, such as *what is dark matter?*, theorists need to take their time. There's no rush. Just get it right.

With exams, such as those faced by *Worried Lad*, I would start them at ten in the morning and finish them at five in the afternoon. People could come in at any time and leave at any time, but only the once. Moreover, I would allow them to bring in whatever books and notes that they wanted. Exam questions should not be so much about memory work (although there is a case for knowing your business) but more about applying *principles*. I think that many students who thought of themselves as exam failures would get much better results given that sort of exam protocol. Moreover, that would be closer to the way science is done.

## Rubric

A *rubric* is the set of instructions at the start of an examination paper, such as telling you how much time you have, how many questions you should answer, and so on.

Usually, the rubric in our Maths Department exams included the instruction "*Do not start until told to do so.*"

Imagine the surprise one year when a typo (typographical error) appeared in one exam's rubric. It read

"*Do not start until to old to do so.*"

After that typo, I often imagined walking past an exam room full of skeletons sitting at their desks, waiting to start their exam.

On the subject of exams, I suppose they're inevitable. If you have to sit a written exam, you should plan carefully what you do. Here's two anecdotes that illustrate different aspects of this advice.

## Mister Surfboard's Choice

There I was, invigilating a third-year Finals exam in mathematical physics. It was for fifteen credits, so the exam rubric was to answer four out of six questions over two and a half hours. It was a sunny Summer's day and it was going to be painfully boring as usual. Moderating an exam in those days was done by all academic staff, and virtually no one could get out of doing it.

The exam started and the clock began its dull *tick* ... *tick* ... for what would seem like an eternity. The students opened their answer books and the race was on.

In all my years of invigilation, I never caught anyone cheating. I never even had a suspicion that someone was cheating. Perhaps I was fortunate, because whenever someone was caught cheating (it did happen once in a blue moon), there was an infernally complicated procedure to go through.

Half way, one of the students suddenly stopped writing, closed his answer book, got up, and left. That was in order. After half-time, students could leave if they had had enough. But they could not come back, once they had left.

When he had left, I went to that student's desk and picked up his answer paper. I needed to write on the cover the time he had left. That was the rule, in case of some subsequent enquiry. Casually, I opened his answer book to see what he had written. Question One looked in order. The student seemed to know his mathematical physics on that topic. I turned the pages onto Question Two. Same thing, reasonable results.

Turning to what should have been the third question answered, I was confronted by something completely unexpected. A *soliloquy.*

The student had written out two pages of words that had nothing to do with the exam directly. It was a long discourse about the sort of life he had planned for himself after his finals had finished. I really wish I had taken a smart phone photo of those two pages, but smart phones were not even on the planning board at that time. The student's soliloquy started something like this:

"*Ah well, that's me finished with exams. I've worked out that if I get these first two questions perfectly right, I'll get at least fifty percent on this exam. Given how I've done on my other exams, I reckon I should get the 2-2* (Lower-Second Class degree classification in the UK system) *that I need to get that accounts clerk job in Skegness* (a popular sea-side town in the UK), *allowing me to go surfing every weekend in the North Sea...*"

On and on he went, writing over two pages about this idyllic life of leisure he was going to have.

I was outraged. How dare he treat our exams in this way! Restraining my anger, I left his answer book on his desk and resumed invigilation.

A day or so later, I collected all the scripts from that exam from the Examinations Office, took them home and started to mark them. By that time, I had calmed down about *Mister Surfboard*, as I started to think of that student. It was particularly important not to harbour any anger when marking any scripts, let alone his. That's one of the responsibilities that you really do learn in Academia. Over the years, I learnt not to be irritated with the vagaries of student's handwriting, such as tiny letters requiring an electron microscope to read them, or the sometimes offensive doodles students occasionally sketch in during exams. I started marking his script as objectively and professionally as I had to be.

And it was good. His two answered questions showed that he was no dummy. By the time I had marked those two questions and was once again reading his soliloquy, I had a new opinion of Mister Surfboard. *It was quite in order, what he had done.* He had indeed gotten close (not entirely, but quite close) to full marks on those two questions, so he did not in fact fail that exam. Actually, I was impressed. I realized that it had been his *choice*. He had decided to go for the minimum needed to get his 2-2 degree. Why should I look down on his choice in life?

When came the examiner's meeting some weeks later, I saw that indeed Mister Surfboard had passed at the 2-2 classification level. I realized that the world contains all sorts of people. Some of them make decisions that we might not understand or agree with. If those decisions affect only themselves and harm no one else, who are we to judge them adversely?

Here's another real-life anecdote about making the wrong sort of exam plans.

**The Film**

Now *Film Boy* was never an outstanding student, at least as far as his exam results were concerned. I suppose not everybody has the right technique for doing well in exams. But revision and preparation really are needed if you want to pass at all. You just have to train yourself to pass an exam, just the same as you have to train yourself before a race, even if you don't expect to win.

The problem was, it became clear in April of the year in question that *Film Boy* was not preparing. In fact, he was doing no work at all. As his personal tutor, it was my job to keep an eye on him and my other personal tutees. Lecturers who were giving out problem sheets and marking them in the run up to the Summer exams would report back to all colleagues how the students were doing in their problem sheets. Indeed, whether they were attending the Workshops and Problem classes at all.

That's how I became aware that *Film Boy* seemed to have dropped out of University. No work, no attendance. So I called him in by email.

That was hard work. He did not reply. I sent another email, then another, all with no reply. So finally, I sent an official, scary-looking letter with our University Logo on it (so it looked official) requiring him to see me immediately.

Finally and reluctantly, he came. I could feel the resentment flowing from him over me, as if I was wasting his time. He sat down and I started to try to find out what was going on. He seemed in good health and was his normal brusque, irritating smug self. Eventually, after some pressure, he told me what he was up to.

"*I'm making a film.*" he announced, as if that was the answer to life and the Universe.

It seemed that, a couple of months before his very last exam, on which his subsequent career might well depend, *Film Boy* had decided that making an amateur film about some half-baked topic was more important than anything else. It was preoccupying him night and day.

The topic was so trivial I cannot remember what it was about. It was not about helping humanity or anything like that. I would have remembered it if it had been. It was not even a chance to get into the film industry by showing what he could do in that respect. If it had been, I might have understood his sense of priority. No. I don't remember what it was about, but I do remember being sure that it was a useless, meaningless topic and that he was wasting his time at a crucial point in his life.

I put the case to him that the day *after* his final exam, he could work on his film, night and day, to his heart's content. No. That was unacceptable to him. He just *had* to do it there and then.

He went away and I did not see him again, even after the exams were over and done with. He did pass his Finals, but not as well as he could have. Unlike most students, he did not come to see me on his last day at University in order to get his results. He had them posted to him. As far as I know, he's still making his film.

There is a sting in the tail of this anecdote. About a year later, I got a phone call from the manager of a Temping Agency, asking for a verbal reference for *Film Boy*. It seemed he had not settled down to any permanent career and was still floating around looking for a job.

As a personal tutor, I was supposed to write references for my tutees. Usually, it was a very great pleasure. When a personal tutee of yours has come out with a First Class degree in mathematical physics, having been a super student over the previous three or four years, giving your best possible reference for them to future employers is actually a privilege. I've had many such great students.

But, over the years, there were one or two students such as *Film Boy* for whom any sort of glowing reference would be dishonest and, possibly, liable for legal action. What could or should I do? Could I tell the manager of that Temping Agency over the phone that *Film Boy* was a layabout, a misguided fool, and someone who could not adhere to their academic duties?

I decided long ago that I would never give bad references, except if there was concrete evidence that a student had been violent or criminal. Fortunately, that had never been the case with any of my personal tutees. I would always give a great reference when I believed that a student deserved one, even if that student had been rude to me personally. But a bad reference? What does that mean? Could I know exactly what was in *Film Boy*'s mind when he decided not to study for his exams? Perhaps he was showing commitment to some friend and was helping them out, at the cost of his own studies. That would be something I could not know about. Besides, deciding how to study for exams is entirely a student's choice.

So, when asked, I said nothing about *Film Boy*'s attitude to his exams that previous year. I gave a factual response, stating the actual marks he had obtained in his exams, with no personal opinion thrown in.

The Manager pressed for more details. I hesitated, and then repeated the factual information that *Film Boy* had been a student in our Department and that he had got his degree at the 2-2 level with the following marks on the following exams. Those were the facts. Anything else would have been subjective on my part.

The Manager hesitated for quite a while, as if he was absorbing the implication of what I had said. Then he said that he understood, thanked me, and rang off. I got the impression that he had guessed that I had given the most positive reference for *Film Boy* that in all honesty I could.

I imagine there are people who read this who will disagree and say that I should have slagged *Film Boy* off. Over the years, I've thought about this a lot. I'm not God. Who am I to judge a person's motives? When it comes down to it, I think the only fair reference system is based on cold hard facts, not opinions about how "good" or "bad" a person is. After all, *Film Boy* had got a 2-2. That was not a bad result, although he might have done better without his film getting in the way. But that was his choice, after I had given him my best advice that it was a mistake to make his film right at that time.

## 13.7 Just do it

The above anecdote may well find resonance with many readers facing exams in a few month's time. I remember how difficult it was in the run up to my final exams, trying to focus my mind on those exams, whilst all the time I wanted to read about exotic topics such as quarks and black holes.

There is a saying, or mantra, that I employ now whenever I don't want to do something mind-bendingly tedious that simply has to be done, such as a bit more typing on my book. You may find it helpful to adapt it to your situation. Whenever I wake up at five in the morning on a cold winter's day and I want to stay in bed, rather than get up, go to my computer, and do an hour's typing, I just say this mantra: *The book won't write itself.* That *always* does the job. Right there and then, I'm reminding myself that no one else in this entire universe is going to come to my assistance and write that book for me. It's down to me and only me. That reminder of cold reality always acts on me as if someone had just thrown a bucket of freezing water over me. I get up immediately and just do what has to be done. Likewise, if you've got an exam coming but you want to go out with friends or play a computer game, just say the magic words: *that exam won't pass itself*, and start revising.

Here's a joke on the philosophy of taking direct action and getting things done.

**The Philosophy of Direct Action**

The link between *living* and *doing* was recognized in antiquity and much debated.

The great philosopher Plato said "*To be is to do*". By this, he meant that the mere process of living a life was all the action that a person needed to take.

His pupil, Aristotle disagreed. He said: "*No. To do is to be*". By this, he meant that living a life is meaningless without doing things.

For over two thousand years, philosophers debated the wisdom of these great words. No one could improve on them, until the philosopher Francis Albert Sinatra gave a much deeper insight. He said

"*Do be do be do.*"

## 13.8 Exam Technique

Sitting a written exam is a dreadful experience, but I guess there's no better system really. Oral exams, where an examiner asks a candidate direct questions and listens to their answers in real time, reveal important details about that candidate, such as how confident they are. It's probably quite hard to cheat in such an exam.

There is an obvious problem with oral exams. They are ephemeral, ghost-like performances that vanish the moment the candidate has left the room. The great advantage written exams have is that the evidence stays around, sometimes for years. My Maths Department had secure rooms full of exam scripts written several years previously. Those scripts were available in case of any enquiry or need to remark.

The chances are that you will sit many written exams before you're through. Here are a few pointers about exams that I found helpful to me in my time as a student.

### *Sleep*

Don't revise late into the night before an exam. Plan your days and nights well before exams start. Make sure your bedroom is quiet, at a correct temperature (cooler is best), has good ventilation, and has no electronic devices in it such as a television, computer, tablet, or mobile phone. Remove such items if any are there. A particular modern issue is that the blue light from mobile phones can suppress melatonin production, affecting the circadian rhythm that governs our day and night activities.

### *Revision*

Plan revision, but if you are late with it, don't panic. It's better to get good sleep and rely on your wits in an exam rather than revise too much and rely on memory. Here's what happened to me in the most important exam I had before going to University, an S-level (Special Paper) in Physics.

## Lyman, Balmer, . . .

At school, I had chosen to do three Advanced level courses in the run up to going to University: Pure Mathematics, Physics, and Applied Mathematics. You may have read in Chapter 1 how three months into the two year course, I had changed from Chemistry to Applied Mathematics A-level. In addition to my three Advanced Level courses, I opted to sit an additional exam: the Physics Special (S)-level. By that stage I knew I wanted to be a theoretical scientist.

S-levels were meant to be tough. There were six questions and you had to do four in a three hour exam. Over the months in the run up to the S-Level exam, I revised thoroughly in Physics. In those days, the Physics syllabus was quite sophisticated. There was a lot of abstract theory to cover, meaning that formulae were usually expressed in symbolic, algebraic form rather than meaningless numerical expressions.

I think that approach to theory had a great impact on my subsequent attitude to mathematical physics. There's a world of difference between a question that asks you to calculate the distance a particle will drop vertically from rest in time $t$ under a gravitational acceleration of $g$, compared with a question that asks you to calculate how far a particle will drop vertically in ten seconds under a gravitational acceleration of 10 metres per second squared. The first question has answer $\frac{1}{2}gt^2$, which tells you a great deal, whereas the second question has answer "*five hundred metres*" which tells you nothing more.

As I revised, I came across Bohr's 1913 theory of the hydrogen atom. It's a wonderful theory if you know a little Newtonian mechanics. I was hooked. I read all about the theory, including the equations of motion and the assumptions that Bohr used. In particular, I memorized the names of the first five lines of the hydrogen emission spectrum (named after the experimentalists who had observed them first): *Lyman, Balmer, Paschen, Brackett,* and *Pfund.* Good old *Pfund.* Remembered for all time as the man who observed the fifth line in the hydrogen atom spectrum (1924, long after Bohr's model came out). Well, I certainly remembered him. Or did I?

The Advanced level examinations had finished, and now it was the turn of the S-levels. The day of my S-level exam came. I left my class and walked towards the examination hall, quite alone and rather apprehensive, as if I was going to an execution. No one else was doing any S-level.

It's strange how apparently small incidents can have a significant effect on you. As I came to the door of the examination hall, I passed a teacher who had never taught me in all the six years I had been in that school, and who had never spoken to me before. We had, essentially, occupied disjoint subsets of that school in space and time, until that moment. He saw me, and out of the blue, wished me *good luck*. He obviously knew what I was going to attempt. The realization of that fact electrified me. Suddenly, I felt a surge of confidence washing over me and I entered that examination room eager to do battle.

I turned the paper over. It was a gift, because one of the six questions was "*Write an essay on any topic in physics.*" The Bohr atom was going to be my essay.

I started on my essay almost immediately. I have always believed that, in any subject, its origins and history are fundamental aspects of it that you neglect at your peril. Yes, you may well learn the mathematics of Einstein's General Relativity without the history, but suppose then you decide to work on a variant of Einstein's theory, a variant that you thought was original. There were indeed several mathematical physicists such as Gunnar Nordström, Gustav Mie, and Max Abraham who developed related theories of gravitation at around the same time that Einstein was developing General Relativity. It's important to know which of their ideas were empirically good and which had flaws. Otherwise, you might find yourself spending months covering the same fruitless ground as those pioneers.

I started my historical account of the development of Bohr's theory with a review of the hydrogen atom spectrum. "*Blah blah blah . . . and the first five lines of the hydrogen spectrum are named after their discoverers: Lyman, Balmer, . . .*"

I could not remember Paschen, Brackett, or even good old Pfund. I was stuck!

Then the dreaded thought came. The exam marker would see that I was working from memory, nothing more. S-levels were supposed to be about *understanding Physics*, not about reciting history. What to do?

In panic, I decided to leave that question. I would not abandon it. Like the famous General, I would return!

I left three or four blank pages after the page where Lyman and Balmer were waiting to be joined by Paschen, Brackett and good old Pfund, and started the next question of choice.

It was a three hour exam. Two hours after my debacle with the Bohr atom, I was ready to have another go. I went back to Lyman and Balmer, and lo and behold: Paschen, Brackett, and good old Pfund, suddenly joined them.

At that instant, all came flashing back and I went on to complete that question fully, giving a comprehensive historical account with all the relevant mathematical equations worked out in detail.

The critical lesson I learnt in that exam was that subconscious processes exist and are at least as important as our conscious processes. Years later, I came across scientific evidence that consciousness is really an afterthought, the icing put on a cake after it has been baked. When I had left that Bohr atom question unfinished, my conscious mind could do no more, right there and then. But that was not the end of the story. My subconscious had not given up. It continued to work on it, eventually retrieving the memories I had laid down in the run up to the exams.

### *Motivation*

I don't know whether scientists have a good theory yet behind all the brain's processes, but more and more research is being done on the brain and its functioning. For instance, the link between motivation and memory has been looked at [Murty and Dickerson (2016)].

Motivation undoubtedly plays a role in creativity, so this should be of interest to anyone either teaching mathematical physics or thinking of going into it. Psychologists have classified motivation into two types: *extrinsic motivation* and *intrinsic motivation*. The former describes the motivation of a person who is concerned what other people think of their work, on their own reputation, and the prizes that could be gained (such as a Nobel Prize). On the other hand, the latter describes the motivation of a person who wants to do their work for its own sake.

These two forms of motivation have been studied, for instance, in *creative writing* by Amabile, [Amabile (1983)], who concluded that

"*To the extent that parents, teachers, and work supervisors model and express approval of intrinsic motivational statements about work, intrinsic orientations and creativity should be fostered. By contrast, to the extent that extrinsic statements are modelled and extrinsic constraints on work are made salient, extrinsic orientations should be fostered and creativity should be undermined.*"

My view is that extrinsic motivation is shallow and potentially dangerous, in that lack of success in extrinsically motivated work may catalyse feelings of failure, rejection, and worthlessness, leading to depression and worse. Recognizing the danger in this respect, I took care many years ago to try to increase my intrinsic motivation to be a mathematical physicist and reduce the natural extrinsic motivation a person invariably has. It's only human to want to be recognized and applauded. But taken to excess, that ambition can be destructive. The best policy, in my humble opinion, is to do your work as best you can, because you find it fascinating for its own sake. Whatever merits that work has, good or bad, will get recognized as such sooner or later, perhaps long after you've passed away.

### *It's not all about exams*

There's too much pressure these days on exam performance. I pity the student who gets the best grades but doesn't really understand the principles of any subject. I've seen quite a few students like that over the years.

Yes, exams are important, but only up to a point. Unfortunately, many students are driven solely by obtaining the best exam grades that they can, regardless of personal motivation, perhaps because of parental pressure.

## You're not allowed to hit students

One of the saddest questions I was ever asked during a lecture was thrown at me a few years ago, five minutes into the very first lecture (out of a total of twenty four) of a third year module (lecture course) on General Relativity. Following my policy of trying to motivate the students, I started by going into the history of the subject. That is a great story and it's one that is still being written by many scientists. As I've explained before, knowing about the origins of a subject seems to me to be an indispensable part of any subject.

A student sitting at the front of that class of thirty put up his hand and asked "*Is this examinable*?"

That was one of the very few occasions that I came close to losing my job.

The point here is that students should not see their studies in terms of exams alone. That is ultimately a counter-productive point of view. Think of your mathematics and physics courses as explorations of the human mind, of what other people have thought, and, ultimately, of what you're capable of thinking. You will only have a limited number of years to appreciate, develop, and apply the immense potential of your mind, and you won't do that if you don't have some feelings of affection for your subject.

# Chapter 14

# Tools of the Trade

*The craftsman who wishes to do his work well must first sharpen his tools.*

*Analects of Confucius* [Eno (2015)]

This chapter may come across as tedious and stating many obvious points. However, I could not avoid including it in this book. Whilst to me being a mathematical physicist is more of a way of life rather than just a job, it's useful to think of it from a practical point of view. I mean, a plumber who didn't have a set of tools would be helpless, and an artist who had no paint would be hopeless. It's no different being a mathematical physicist. The conceptual problems that you deal with in mathematical physics are hard enough: you don't want to be distracted by relatively inessential factors such as having nothing to write with, or running out of paper. You simply have to gear up properly *before* you start theorizing. Here are a few pointers that I have found useful in my time.

## 14.1 To write or to type

The modern student will think that whether you write by hand or type on a computer or tablet is a no-brainer of a question. The answer appears to be obvious: typing on electronic media has many more advantages, such as ready backup and not needing paper, binders, and pens. If you think about it a bit, why bother typing at all, when you can download everything you need from the internet?

There are several reasons why I believe handwriting is preferable. I'll list pros and cons presently. Meanwhile, here's an anecdote on the theme of writing versus typing.

### Cut and Paste

Working in a maths department meant I had a range of teaching duties. One of these was supervising Final Year projects. Such a project would occupy most of a semester (half of the academic year) and contribute a substantial number of marks towards the overall final assessment of a student's degree.

Project work is fundamental to a budding mathematical physicist's development. Any half-decent project will have within its parameters some opportunity for independent thinking and the advancement of knowledge, even if by only a modest amount. Project work requires commitment, the development of good research habits, focus on a given problem, and finally, the acquisition of write-up skills so necessary in communicating results.

Concerning the particular project I'm writing about, it was one that I had devised and it had been entered into the departmental handbook. Final Year students could read about our projects and decide which one they wanted. Then they would find a supervisor and be guided by them over several months into developing some sort of theory about the topic. Then they would try to apply the theory to a problem and finally write about what they had done. There's precious little difference between doing all that and doing actual front-line research, except perhaps for the depth that a student can get to in a few months. An important aspect of those particular projects was that a student did it on their own, not in a group, making it excellent training for mathematical physics.

The semester started but my assigned student did not show up for the first meeting.

Fine, that was his choice. I duly emailed him and he replied that he would come next week.

Next week, the same thing happened. He didn't come then either. In fact he didn't come for a whole month.

Finally he came, quite apologetic. He had been abroad on some religious trip.

Over the years, I learned not to get worried about such matters, but at the time I was concerned. Some things such as religious observance are essential to many people. These days, Universities are much more aware of this than they used to be and make all sorts of allowances, but in those days that was not always the case. My worry then was the time that he had left himself for his project was too short. However, on reflection, I realized that this was up to him, his choice. The University had, after all, given formal permission for his absence, but I had not been told at the time. When I thought about the issue a bit, I concluded that as long as the project was completed and his write-up handed in on time, what would be the problem?

Time turned out not to be the problem. The real problem was this. When eventually he came, he seemed pleased with himself. He told me that he had indeed done some "research" over the month he was away and he was going to show me there and then. He then put his laptop on my desk and booted it up, saying that he had downloaded some useful articles from the Internet and was going to show them to me.

There was nothing else. He had had no thoughts himself about the project, had taken no notes whatsoever, and had only downloaded files. That was all he though "research" actually was: the finding of other people's articles and the downloading of them. He had not absorbed any information from them, because when I asked him, he couldn't say anything about those articles. He had not actually written or typed anything himself.

I asked him to switch his machine off, saying that I could download those articles myself. I quizzed him on what he had actually learned from those downloads, and what was he going to do with that knowledge.

There was no answer. He seemed puzzled, until I explained to him that real research was not just finding out what other people had written, but understanding what they had written, and then applying and developing that knowledge to tackle some problem in a new way.

I asked him to come next week without any laptop or electronic device, but only with actual, real paper, on which he had written his notes and thoughts, by hand, himself, in pen or pencil. I was not even worried as to whether he had copied onto paper, word for word, any article he had seen online. The mere writing out of other people's ideas would have some sort of impact on him.

I was delighted to see that when next he came, and every week thereafter, he had done exactly as I had asked. He presented only hand written notes, correlating details from various articles he had read. He started to develop his ideas on paper, doing calculations and working results out for himself, all of which contributed towards a successful project.

The moral of this anecdote is that modern electronic technology is good and has its uses, but cannot substitute for the ancient technology of hand, eye, brain, pen and paper, used by the human mathematical physicist. If you want to become an independently thinking mathematical physicist, my best advice is to use electronic means when it's needed but stick to good old fashioned writing by hand as much as you can.

There are two points here concerning electronic versus analogue, or machine versus human, if you like.

My first point is that there is a growing concern amongst some scientists, such as Stephen Hawking, that humanity may be threatened by artificial intelligence (AI). I think there is something to that, given the way the Internet is gathering data about us individuals. I don't know whether machines will ever replace the human mathematical physicist when it comes to creativity, for that comes from the depths of the human subconscious, a place where no machine has gone to, yet.

I suspect mathematical physicists will never be replaced entirely in this respect, but I could be wrong. The brain is not structured entirely by genetics but also by *experience*. What happens to us, including deliberate conditioning processes, reinforces the way neural networks are patterned into our brains. Although currently rather primitive, *machine learning* based on artificial *neural networks* suggests that machines can learn complex behaviour by experience. This is currently of commercial importance in the the development of self-driven vehicles.

It would be contrary to *Nullius in Verba* to assert that machines could never replace humans. Humans are, for all practical purposes, biological machines. Replacing the adjective *biological* with *electronic* may be just a matter of scale. The human brain has about eighty six billion neurons [Herculano-Houzel (2009)] and they can be interconnected in a vast number of patterns. It may be possible to construct electronic or even biologically based machines with similar capabilities. In a way, that's what a baby is, after all.

My second point is that recent research suggests that handwriting increases critical thinking and promotes long-term memory recall [Mangen and Velay

(2010)]. I think there is a lot to that. Typing quickly does not mean you are necessarily focusing on the *significance* of what you are typing. You may be focusing on the *accuracy* of what you are typing, but that is just one aspect of learning a new subject. Indeed, a good modern spell checker can even allow you to be pretty dismal at spelling. On the other hand, pausing to think a second or two before you write down the next word may well give you an opportunity to rethink your words and possibly, trigger new thoughts along a new mental pathway.

## 14.2 Pens, pencils, and erasers

I'm pretty sure most people have been in a situation where they have to sign something, or leave a brief note for someone, and there's no pen or pencil handy. There's fewer things more frustrating than that. An even more frustrating situation for a mathematical physicist is when you're in a new place, such as on a train, and you have a sudden urge to write down your thoughts (which at the time you're convinced will unlock the secrets of the Universe), but you have nothing to write with.

I solved that particular problem some years ago by going on the internet, looking at the reviews, and then buying several job lots of pencils and gel pens. I don't mean just a few. I bought boxes of fifty pencils and pens at a time. I still have some of them ready for use. I scattered them generously around the places where I tend to sit: in my home office, in my car, next to the television, next to my bed, my mobile office (see below) and so on. What you'll find is that unless you're careful, those pens will start to disappear quite quickly, as other people will find them useful too. The point is, don't solve the pen problem by constantly looking for the only one you have. Get as many as you can so that at least one is close when you need it.

There's an important question here concerning the actual writing technology: *pen or pencil*? Here are my experiences.

### *Pencils*

Pencil technology is far more subtle than it first appears. A pencil is pencil, isn't it?

No! Pencils come in a variety of grades that can make or break your writing experience. Believe me, if you're doing a lot of calculations in pencil, choice of grade can become a significant factor in how enjoyable it is. In Europe, where I live, pencils tend to get sold on an "HB" scale. The *H* refers to the degree of *hardness* of the pencil lead[1] whilst the *B* refers to its degree of *blackness* on paper. What a pencil deposits on paper is a mixture of graphite (a form of carbon) and clay. It's the proportions of the two that determine the grade of the pencil.

After many years of use, I eventually settled on the HB2 standard of pencil, as it is often referred to in Europe. Out of all available pencils, I found that the HB2 standard pencil gives the best results in the long term. Here are the pros and cons of HB2 pencils.

**Pencil pros**

• HB2 is soft enough to erase well without lifting the paper surface, unless you do it too often at the same place.
• HB2 does not smudge easily.
• HB2 scans relatively well. I found that calculations I did on softer pencils do not scan well. Some of my archives from several decades ago, before I appreciated this aspect, are almost invisible even under the darkest scan that I have available, so they're effectively lost unless I go over them with a harder pencil or pen.
• There's something about writing with a good pencil on good paper that beats other technology. It can be a sensuous experience. With a HB2 pencil you can shade gently or embolden firmly, something you cannot do with pen or ball-point.
• For some reason, my handwriting seems better to me when I write in pencil than in any other medium such as gel-pen or ink pen. I suspect it may have to do with the greater care you have to take with a pencil in order not to break the point. Slowing down your writing is not a limitation, in my experience. Quality of writing will always beat quantity of writing.

**Pencil cons**

• You have to have an eraser handy to take advantage of the full potential of a pencil. I bought one job lot of fifty pencils without attached individual erasers. That was a serious mistake which I got around by buying an *electric eraser*. It's like a fat pen that you hold. When you switch it on (it runs on a standard AA battery), the eraser tip vibrates gently

and clears away HB2 marks efficiently (depending on the paper surface of course).
• You have to sharpen pencils, which is probably their biggest disadvantage. After finding electric pencil sharpeners were too vicious, reducing long pencils to short stubs in seconds, I invested in a really neat manual pencil sharpener that sharpens *so much*, and then stops sharpening once the point is sharp. It's a really clever piece of metalwork and is highly recommended. You can find such devices through the internet, but good ones are not cheap.

If you go for pencil, don't go for cheap options. A wise person I know once told me that "*A poor person cannot afford to buy cheap*". There's a lot of sense there. I once bought an inexpensive job lot of fifty HB2 pencils from an unbranded manufacturer. Each pencil had its own eraser on the blunt end. But after six months, those erasers were rock solid and unusable. Meanwhile, more expensive quality pencils that I had bought months before and stored in exactly the same place had erasers that were as soft as if they were new. Even now, several *years* later, those quality erasers do the job perfectly.

A good variant of the traditional wood pencil is the mechanical pencil. Here too there is great variety in grade and quality. After some online research I found a brand that has stood the test of time and bought a job lot of ten. They're not cheap, so I keep an eye on them. They come with an inbuilt eraser at one end, so that takes care of that issue. I found that the 0.5 mm thick HB2 lead was best for mathematical physics calculations done with such a pencil. Too fine a lead means it breaks constantly, whilst too wide a lead means that your symbols such as superscripts and subscripts tend to get confusing. This is *not* a trivial point in mathematical physics. If you are doing any tensor calculus, as you may well do if you are going into any relativity or quantum field theory, you'll be encountering superscripts and subscripts all the time.

### *Pens*

Pencils have their uses but occasionally it's better to use pens. There are three classes of pen I recognize: *fountain pens*, *gel and ball-point pens*, and *technical drawing pens*. I'll discuss these separately.

### *Fountain pens*

I don't use fountain pens at all, as they have too many disadvantages:

**Fountain pen cons**
• They can leak.
• It's messy refilling them.
• If they have dried out, it's messy to wash them out.
• Only fine-tipped nibs are suitable for mathematical work, in my experience, but such nibs are too fragile for extensive use and get clogged up too easily.
• Ordinary fountain pen ink dries too slowly in my experience.

### *Gel and ball-point pens*

**Pros**
• There's no need to sharpen such pens and they have instant applicability. Such pens tend not to dry out.
• A good gel pen is a delight to use.
• Black gel pens invariably give good scan results. I would not use any other colour if I anticipated scanning my calculations.
• Such pens are best suited for rough, preliminary calculations or first draughts of research notes or articles. These are best done on recycled sheets.

**Cons**
• If you buy a cheap job lot of ball-point pens, as I did once, you may find a large percentage simply don't work. It's obvious really, once you realize that a ball-point pen requires a *ball* to work properly. If the ball in such a pen is not well machined, it will simply not work. So, go for quality rather than quantity.
• For some reason, my handwriting is worse with such pens. I think it has something to do with the speed of writing, which can be too fast.
• Before you know it, you will have a collection of *empty* gel and ball-point pens, as this type of pen is not economical to refill.

### *Technical drawing pens*

Early on in my research career, I rediscovered the great joy and value of *technical drawing pens* based on *drawing ink*. That is a soot-based, exceptionally black, non-toxic ink that is permanent and scans beautifully. Such ink was used thousand of years ago in China and India. I first came across such pens and drawing ink at school, when I did a technical drawing course.

When I started my PhD (doctorate) some years later, I quickly became disillusioned with ordinary fountain pens, which is why I always avoid them. They leak, the nibs wear out, and ordinary pen ink has several disadvantages, such as smudging and blotting. I remembered my old school technical drawing pen, found it, washed it out, refilled it, and immediately discovered that it is simply superb for mathematical physics calculations. With such a pen and the right paper (see below), you will find that doing mathematical physics becomes a hugely satisfying experience. Here are the pros and cons of such a pen.

**Drawing ink technical pen pros**

• By design, technical pen give consistent lines. There are no variations in stroke width such as you would get with a broad nib fountain pen. This is critical in mathematical physics, where superscripts and subscripts are constantly encountered. I found that a 0.5 mm diameter point suits me perfectly in this respect

• The results are clear and scan brilliantly.

• The results do not fade, even over decades. Documents I wrote over thirty years ago with a technical pen look as if I had done them yesterday.

• If you buy a large inexpensive bottle of drawing ink, you can refill your technical pen many times, so the original high cost is low over many years.

• I use my favourite 0.5 mm technical pen for writing up final versions of calculations. Because of the nature of the pen, I write slowly and carefully, and the results are generally immensely satisfying.

• Such pens are not cheap, but just one can last you a lifetime, without a change of nib. The reason is that you can refill such pens easily yourself and, more to the point, the nib technology is different to standard fountain pen nib technology. In a technical drawing pen, the nib is actually a hollow tube through which there is a really fine wire that can move in and out slightly. As the nib is touched on the surface of paper, the inner wire is pressed into the body of the pen, and drawing ink starts to flow out. This

means you don't need to press hard to get an excellent flow of ink. Such nibs tend not to wear out in my experience.
• Provided you are careful with maintaining your technical pen, you will not get ink blots over your work.

**Drawing ink technical pen cons**
• Nothing is cost-free. Such pens are not cheap initially, but in the long run are really economical.
• You do need to clean such a pen occasionally. Washing out with water is straightforward, but do be careful, as the ink will spoil your clothes if you are careless.
• You have to allow the ink to dry for a few moments before you put anything on the page you've just finished. In my experience, it dries faster than ordinary fountain pen ink.

## 14.3 Electronic handwriting tablets

Here I am referring to devices that are *not* the all-singing, internet friendly devices such as Ipads, Surfaces, and Galaxies that the word "tablet" suggests. The tablets I mean are dedicated to one purpose alone, which is *electronic handwriting*. That sounds self-contradictory, but it means that the writer uses a stylus to hand-write over a touch-sensitive screen and the results are recorded electronically by the device.

Such devices represent an important step towards the eagerly anticipated and long-hoped-for development of *electronic paper*. Technology is not there yet in this respect but what has been developed so far is impressive. People have long imagined handwriting on what looks and feels like paper, making the inevitable mistake, and then erasing that mistake just by pointing to the error with a stylus.

Recently I acquired a remarkable device that allows me to write pages of notes and calculations in fine detail by hand and then export them electronically to my computer as pdf files, without further scanning. This sort of device may well become the medium of choice for the working mathematical physicist, replacing paper for rough calculations eventually. Here are the pros and cons that I have experienced.

**Electronic handwriting tablet pros**

• The experience of handwriting on such a tablet can be quite close to that of writing on paper, depending of the surface of the screen. Conventional tablet surfaces, made of glass, are not as good as surfaces with a designed degree of friction between them and the stylus. The machine that I bought is really satisfying in this respect.
• There is no practical limit to the number of pages you can write on.
• The contrast (degree of whiteness versus blackness on the screen) is good and getting close to the contrast of white paper and black ink.
• They are transportable.
• They can help eliminate or reduce *paperbergs* (see the next chapter for an explanation of what I mean by this term).

**Electronic handwriting tablet cons**

• Such tablets are still quite expensive.
• Such tablets need to be recharged regularly.
• You can have only one screen visible at a time, unless you get two or more such devices. Primitive clay tablets and A4 sheets of paper don't have that limitation.
• One of the biggest negatives usually tends to be inadequate software. The machine I bought is good except for one point which makes its long term use for a mathematical physics calculation problematical: the current version of the software does not allow you to copy anything you wrote on one page to another page. If asked, I would say that this is *the* critical issue that would have to be resolved before I could recommend it to a mathematical physicist. In all other respects, the software and the concept is just brilliant.
• A significant issue can be low *latency*, meaning an unacceptably slow response of the device to accepts and register the actions of the stylus.
• Glass-based screens are frequently criticized as having too much glare and having inadequate or poor quality backlighting.
• If your device backs up your work to the internet cloud, you will be linked to the fortunes of the manufacturer. If they go bust, you are at risk of losing data. All the issues to do with hacking and security go along with this technology.

## 14.4 Paper

I turn now to another critical factor in handwriting documents: the *paper* that you write on.

You may believe that paper is paper, and that any brand and type of paper is as good as any other. In this respect, you would be seriously mistaken. Paper quality can be a critical factor as far as hand writing is concerned. The paper that you write on can make or break your enjoyment of writing. If you don't enjoy it, you will be reluctant to do it.

There are several factors involved in choosing a good paper for mathematical physics calculations.

• **Paper density** is perhaps the most critical feature here. In Europe, paper density is conventionally stated in terms of *grammage*, or the weight in grammes per square metre, denoted $g/m^2$ or equivalently, gsm. The lower the grammage, the flimsier the paper will be. Office paper will generally be rated at 70 or 80 gsm. My experience is that if you can get 90 gsm paper, use it for final versions of your calculations (I mean, the final calculations that you will archive).

• The **surface** of paper is a critical factor that should not be ignored, especially when writing with pencil. A brief look on the internet reveals hundreds of varieties of paper surface available, in a range of tones. I do have a clear personal favourite but for obvious reasons will not advertise it here. It is 90 gsm, so both sides of a sheet can be used without any problem, and is really smooth to write on.

• **Margin, ruled, or plain?** This is a surprisingly important factor if you are doing a long mathematical physics calculation. I have found that *narrow ruled and margin* A4 paper works best for me. Having rules (horizontal lines) means that you can lay out a calculation effectively. With narrow ruled paper, I usually space out equations with two lines to each equation. That gives scope for later insertions and, additionally, gives a good visual impact on the eye. That is an important feature of any hand calculation. Too many cramped symbols can slow down the necessary reviewing that you should always do of your work.

As for the margin, it keeps you away from the edge, which you should reserve for binding your documents in some way, such as putting them in a ring binder or spiral binder notebook.

## 14.5 Whiteboards

If you've ever watched that brilliant series *The Big Bang Theory*, you will be familiar with the *whiteboards* that Sheldon Cooper and Leonard Hofstadter use to do calculations. Years before I saw that series, I experimented with such boards.

They look good, but they have their issues. First of all, they're not very portable and you have to stand up to use them. That gets tiring quite quickly, so before you know it, *you don't want to use them any more.* Doing long mathematical physics calculations requires comfort if it's to be done over extended periods of time.

Another point is the sheer impracticality of whiteboards. I had an enormous one in my Maths department office, covering one whole wall. It ended up being covered with masses of notes, memos, partially completed calculations, and plain rubbish. You would start off doing a detailed calculation in one corner, but end up very quickly running into bits of another calculation started elsewhere on the board. If you tried to erase the old bits, you would lose the information they held, so you had to divert your attention from what you were trying to do at that time and copy out the old bits onto paper before you could rub them out. So in practical terms, it would have been better to use paper in the first place.

The great thing about white boards is that they do look impressive, as in *The Big Bang Theory*, if you cover them with mathematical-looking meaningless scribbles. Casual visitors to your office may well think you're doing something meaningful, even if it's a load of old rubbish[2].

To be fair, there were one or two occasions when talking to my current research student, we would make good use of my white board for a novel calculation. That was only after an intense purge of everything that had been on it beforehand.

## 14.6 Laminated sheets

The search for an effective compromise on the writing front led me to develop my *laminated sheet* technology. A *laminator* is a machine that covers ordinary paper with a heat-sealed transparent plastic covering. Laminators are excellent for preserving fragile documents, such as birth certificates and marriage licenses. I found that if I laminated a clean, white A4 sheet of paper, then it had all the characteristic of a mobile white board, if I used a non-permanent, water based overhead transparency pen. Moreover, such a laminated sheet can be used over and over, countless times. Because lamination makes the paper waterproof, then simply soaking a filled-up sheet in water would clean it completely and instantly. The only issue there would be the need to dry the wet sheets.

I ended up laminating dozens of such sheets, giving me the possibility of doing long mathematical physics calculations. Essentially, I had an analogue of the electronic handwriting tablet I discuss above.

The laminated sheet approach has the drawback of whiteboards, in that you cannot use a document scanner to convert them into electronic form. These days, there is the possibility of using a smart phone to take photographs of your work and using available apps to convert to pdf format. However, that soon gets as tedious as using a whiteboard. Ultimately, I concluded that the best way forward would be the electronic handwriting tablet approach discussed above.

## 14.7 My mobile office

Once you start activities as a theorist, you will find that the concept of a *mobile office* is a really sensible and practical one. What I mean by *mobile office* is not a room on wheels but essentially a multi-compartmental bag full of paper pads, pens, pencils, staplers, rulers, pencil sharpeners, USB sticks, spare glasses (I need them!), and any other essentials needed for doing calculations not only in the field, but actually *in* a real field, should I go into one. My mobile office has a shoulder strap and I can go onto trains, attend conferences, or visit family with it, knowing that I can do my work whenever and wherever I want to. I can also carry a laptop and charger in my mobile office, giving me the facility to do really serious mathematical physics work wherever I go.

## 14.8 Typesetting software

You may have been brought up on Ipads or other Tablets. There's nothing wrong with such things, but to be perfectly honest, I have never found them up to the job of heavy-weight word processing. Sooner or later, you will want to start typing out a big document, perhaps even a book. Your choice of typesetting technology and software will be critical to the experience you will have.

I use a conventional PC, which I've discussed in Chapter 13, *Study, Sleep, and Exams*. The operating system happens to be Windows. I've never used Linux but by all accounts it's a great alternative. The immense (and I mean *immense*) advantage of either of these operating systems is the availability of the superb free typesetting software known as *Latex*.

Latex is the format of choice throughout the scientific community for writing papers, technical documents, scientific and mathematical textbooks, and conference overhead talks. It is *open source*, meaning that there is a vast community of inspired programmers who have developed an astonishing armoury of add-on programs that can be inserted into the basic Latex core program, allowing you the author to write beautifully laid-out articles and books in a huge range of publishing house styles. It's all free, which is as it should be.

All my typeset papers and books have been processed in Latex by myself. A critically important feature is that publishers invariably have Latex style-files that can be used to create final processed versions of your work in whatever *house style* a publisher uses. For example, I'm currently typing out this book in Latex with processed output in the form of a pdf file in World Scientific's $9 \times 6$ book format. It looks just like the book that hopefully you have seen in a bookshop.

I would strongly recommend any budding scientist, mathematician, or mathematical physicist to download available free Latex editors, patiently learn Latex, and produce professional-level documents. There is a steep learning curve, but the results are fully worth the effort. Start slowly and learn as you go. When problems arise, I look online for solutions. In my experience, someone out there will have had the same issues and the experts invariably will have given sound advice that resolved those issues.

## 14.9 Binding and Archiving

It does not take much work to produce dozens of sheets of typed or handwritten paper lying around your desk. Sooner or later, you will want to bind them in some format and archive them. Archiving is really important. I've never thrown away documentation that I have produced, on the principle that destroying paper is irreversible. It's only in recent years, with the purchase of my document scanner and multiple backup hard discs, that I felt comfortable in shredding any documents.

Here's a list of archiving methods that I've tried.

### *Hard-cover notebooks*

Some theorists use hard-cover bound notebooks to record everything by hand. I've got a load of such notebooks and they certainly do the job. Many of them date from my student days. They are more-or-less permanent and keep pages in order.

An understated, important advantage of such notebooks is that they can be stored on shelves as if they were normal books, without falling over. That is not the case with other forms of bound notes.

At one time, I even contemplated hard binding my loose notes into hard-covered form. This would have involved learning the art of *book binding*. Fortunately, I came to my senses: mathematical physicists need time to think, and should not waste it waiting for glue to dry.

There are two downsides to notebooks. First, they're a bit *too* permanent: you cannot re-order the pages or replace a spoiled page with a new one. Second, if you want to scan any pages and store the contents on line, you either have to cut the pages free, essentially destroying the notebook's usefulness, or you have to have a scanner that can scan single pages of books. *That* gets very tedious if you have to scan say thirty pages, one at a time.

### *Ring-binders*

These have several advantage: pages can be easily put in and taken out, and re-ordered or replaced easily. Moreover, separate pages can be inserted in transparent plastic wallets, which can guard against coffee spills (an occupational hazard for mathematical physicists).

A significant disadvantage of ring-binders is that they do not stack well on shelves. Another disadvantage is that large ring-binders are bulky and awkward to use as reference books when doing calculations. Ring-binders are best used in a long term archive capacity, not for frequent access.

### *Spiral and comb binders*

There are various kinds of do-it-yourself binding technologies. I invested in an electric plastic comb binding machine and regularly bind notebooks with several hundred pages of notes.

The advantages are three-fold. First, such notebooks open out conveniently, unlike rind-binders, so that pages can be properly read. Second, pages can be inserted, replaced, and reordered relatively easily. Third, with care, spiral bound notebooks can be housed on shelves.

Such binding technology gives you the nearest thing to creating your own physical books at home, short of learning the ancient art of bookbinding, which involves cutting sheets of paper and waiting for glue to dry.

## 14.10 Desks, chairs, and desk lighting

A good desk is an essential component in any office or home environment where you may have to do your calculations and typing. There is a need to balance two competing forces here. Too small a desk will be uncomfortable, particularly if you need computer monitors (I currently have three flat screen monitors working simultaneously from the same PC), plus whatever you're working on, lying on your desk. On the other hand, too large a desk invites the creation of paperbergs (see next Chapter).

Never ignore the fact that you are a machine with a sensitive skeleton and muscles ready to get tired and sore. Finding the optimal height and location for your desk is a critical factor that will play a fundamental role in your productivity. If for example you do not have a desk but are forced to do your work sitting on the floor or on a sofa, you will be severely hampered.

Hundred of years ago, educated aristocrats who managed their estates personally or were interested in science and literature would often use a *standing desk*, a desk at which a person works standing up. Such desks are still used in various workplaces, such as artists studios. A point advocated by ergonomists in favour of a standing desk is that the user can maintain good posture whereas a normal seated desk invites bad posture. I explored the advantages of such a desk and concluded that it was not for me, because I found standing for more than ten or more minutes was tiring. However, your experience may be better.

The chair that you sit on is as important to productivity as any other factor. A particular hazard is that if your chair is too high, or has insufficient support, pressure can build up on your thighs, with potential risks to circulation.

Given that you have found the best desk and chair for your work, an equally important factor is *lighting*. You can find sensible advice about this online. Good lighting will allow close mathematical work over an hour without adverse effect on your eyes. I invested in a lamp sold by the RNIB (Royal National Institute for the Blind) after consulting online expert advice on this issue. I bought a clamp-on desk light that has adjustable intensity and colour range. The advantage of a clamp-on lamp is that it can be positioned in more places than the usual base-mounted form of light.

Regrettably, these issues may require some degree of investment if you decide to set up a properly equipped base at home from which to do your mathematical physics. Not every mathematical physicist works in a University that provides suitable space and equipment. Having said that, a good mobile office (see above) can allow you to work virtually wherever you want to.

## 14.11 Planning and scheduling

Planning and organizing your work schedules is a critical factor to productivity for a mathematical physicist. The reason is that theories take ages to develop. If you have a bright idea (or so you think), it may be weeks, months, or years before you end up with a finished paper or book that can be sent to a publisher. I have found it necessary to have an overview of the whole process, to ensure I have some control of my time.

In this respect, there is a good anecdote involving the famous mathematical physicist Wolfgang Pauli that I heard or read somewhere.

**Railway Timetables**

In the early days of quantum mechanics, someone once asked the great mathematical physicist Pauli why people still used the laws of physics based on classical mechanics, when they were obviously incorrect compared with the actually observed data. Pauli replied that it's the same issue with the notoriously bad railways timetables at that time: *We need railway timetables in order to know how late the trains are.*

What Pauli meant was that when we do our experiments, it's good to know how far off the data is from our classical expectations.

Given a long term research program (or even study program), it's good practice to plan your daily, weekly, and monthly activities, with a deadline for completion. As you work, comparing what you have actually done and what you had planned to do gives you a realistic perspective. Every day, I boot up my computer and spend the first half an hour recording yesterday's activities and planning today's activities on an hourly basis. I update a graphical display on an EXCEL sheet of my daily activities and have done so for several years.

This may seem to be overdoing things and perhaps a symptom of some compulsive disorder. Far from it. I wish I had started to do this decades ago. I regard planning and scheduling my working activities as a mathematical physicist as a core activity. My graphical display keeps me alert to periods where I have started to slow down, usually due to complacency or getting engaged in some displacement activity. I take care not to be obsessive, so if I've got too bored with mathematical physics (it can happen), I abandon that day's schedule. Too much of that, however, eventually shows up on my long term graph and I take steps to return to normal activity.

I thoroughly recommend any student to find some long term planning method, either using a physical year diary-planner, or a good computer or tablet organizer. One of the key benefits I found in my planning regime is finding out how frequently I was *not* working on mathematical physicist. I never said that becoming a mathematical physicist would be easy or quick. It requires dedication, time, and a lot of work. You can do it, if you have the motivation and plan for it properly.

## 14.12 All you really need

It's all very well me recommending this or that device and advising about expensive desks and chairs, but really, none of them are absolutely essential. Archimedes had no electronics, gel pens, or three-hundred sheet A4 pads of narrow-ruled and margined paper, but that didn't stop him doing great things with his mind. When I think about the technology he had available and what he achieved, I feel quite humbled. Similar comments could be made about Newton and many other mathematical physicists. If you cannot set yourself up with all the fancy equipment I've discussed above, don't worry. You won't be worse off than Einstein in this respect, and look at what he accomplished.

What I'm saying is that there are tools of the trade that you can use to make your work as a mathematical physicist a bit easier, but they are not absolutely essential. If you have a clear mind and the motivation, the proverbial *back of an envelope* plus pencil will be all that you need to do your work, wherever you are, be it in an office, at home, on a train, on top of a mountain, on a beach, in the jungle, or in the desert.

# Chapter 15

# The Mess on Your Desk

*If your desk isn't cluttered, you probably aren't doing your job.*

Harold Geneen [Geneen (1985)]

Perhaps I've convinced you to consider becoming a mathematical physicist. I discuss motivation, writing, conferences, tools of the trade and suchlike in other chapters. In this chapter, I want to discuss some practical aspects of doing mathematical physics on a daily basis.

## 15.1 Paperbergs

You'll discover pretty quickly, when you start in any theoretical discipline such as mathematics or mathematical physics, that you won't be able to work out your great theory just like that. Even simple ideas take a great deal of research, all of which starts to form *paperbergs*. That's my name for the inevitable mounds of unsorted papers blocking up your desk, office floor, garage, and/or loft, that form naturally the more you work. They are the academic equivalent of *fatbergs*, which are unpleasant accumulations of waste products blocking sewers (some fatbergs get to weigh hundreds of tons before they are cleared out). Paperbergs will always form *unless* you take steps to control them.

Many creative people have had paperbergs on their desks, including Einstein. You can see a famous picture of Einstein's desk, taken on the day he died in 1955, that illustrates perfectly what I mean [Tate (2019)].

Is eliminating paperbergs important? There are pros and cons here.

### *Against paperbergs*

There's three reasons I can think of why paperbergs can be a problem.

First and most obviously, they are unsightly and give the casual visitor to your office the impression that you're disorganized. Just imagine going into a lawyer's office hoping for good advice and seeing mounds of papers stacked up everywhere in confusion. Could this person really help you when it looks as if they couldn't organize their own office properly? That's a natural question.

A second, potentially serious problem is with *retrieval*. By this I mean the inevitable need to find some document such as a research paper or calculation that you had done in the dim and distant past. If that document is (perhaps) sitting somewhere in one of three or four paperbergs in your office, you will just have to go through those mounds of unsorted papers hoping to find it. If you have no method of keeping a track of even which *room* that document might be in, you may well find yourself sifting through paperbergs in other rooms as well, such as your garage, or even your loft. This scenario has occurred so often to me that I classify paperbergs as *the* prime target for *decluttering*, the painful task of re-organizing your environment in such a way as to minimize retrieval time. As a working mathematical physicist, you simply cannot afford to waste time on retrieval.

A third, equally serious, problem with paperbergs is that they tend to end up in the wrong places, such as on floors, and that can be a hazard. I've narrowly avoided serious injury on several occasions when tripping over paperbergs that I placed around me on the floor, temporarily. It's surprising how "temporarily" can change into "long term".

### *Don't over do it*

Although having too many paperbergs in your office can be seriously counterproductive, there are pitfalls to being overzealous in clearing them. After all, as you will see from the photograph of Einstein's desk [Tate (2019)], he did not need to declutter his desk to achieve great things.

There's a saying that's frequently attributed erroneously to Einstein: "*If a cluttered desk is a sign of a cluttered mind, what's an empty desk a sign of?*" Regardless of who coined that saying, it's a good question. Here's an anecdote that has stuck in my mind for years, as it gives an important lesson about tidiness.

**The Pens**

For years before this particular incident, I was plagued by all the paperbergs I generated, both at home and in my office in the department. Why could I not keep my work well sorted? Then one day, I realized there was a down side to tidiness.

In any department, you will find a broad spectrum of thinkers. Some colleagues will be constantly active intellectually, while others are just waiting for their pensions. I guess my colleague *Doctor Clutterfree* was one of the latter.

I had never been in *Doctor Clutterfree*'s office before, despite the fact that by the time of this incident, we had worked in the same department for at least ten years. But there had been no reason to go there before.

What brought me to walk into his office that day is the fact that Maths is a relatively unified subject. It has well established conventions and generally, a standard approach to notation. However, mathematical physicists are quite good at inventing their own notation if the current offer is inadequate. A great example is the so-called *Dirac delta*. That's a bit of notation that mathematical physicists often use to represent point particles. It's so weird that when mathematical physicists first started using Dirac deltas in their calculations, many pure mathematicians said they were meaningless rubbish. That is, until about the 1950's, when the mathematician Laurent Schwartz put them on the mathematical map.

In the course of my lecturing duties, I had come across a particular issue involving *integration*. You don't need to know much about integration except it's a clever way to measure the sizes of objects. For instance, suppose you had a straight stick with its two ends labelled $A$ and $B$ respectively. If you wanted to measure the length of that stick, you would put the start of a tape measure at end $A$, roll the tape out to end $B$, and read off its length by the number it gave at $B$. That's quite obvious.

Now suppose you put the start of the tape at end $B$ instead and rolled out the tape to end $A$. What would you get? Obviously, you should get the same answer. That's in line with the principles of geometry, a mathematical discipline, which postulate that the *absolute* or *intrinsic* length of the stick doesn't depend on which order, $A$-to-$B$ or $B$-to-$A$, that you take to measure it.

However, in mathematical physics, the order in which you do things often matters. That's certainly true in quantum mechanics, for instance. In the case of the stick, mathematical physicists might be interested in the *signed distance*, where the direction of measurement does matter. The signed distance *from $A$ to $B$* turns out to be *minus* the signed distance *from $B$ to $A$*. In terms of integral notation, we would write

$$\int_{x_A}^{x_B} dx = -\int_{x_B}^{x_A} dx, \tag{15.1}$$

where $x_A$ and $x_B$ are the coordinates of $A$ and $B$ respectively, for the signed distance along the stick.

Just to ensure consistency with my pure mathematics colleagues, I decided to check with *Doctor Clutterfree* on his perspectives on this notation.

I found *Doctor Clutterfree* at his desk. It was, like the rest of his office, completely clear of paperbergs. Nothing lay out of place. Every part of his office was pristine, including his whiteboard. No great calculation in progress lay on it. I was impressed. Compared with my office, *Doctor Clutterfree*'s place was a showroom of tidiness.

*Doctor Clutterfree* was welcoming. I sat down and started to outline my concerns with the notation. I picked up a whiteboard pen from his desk, went to his whiteboard, and wrote out equation (15.1). I sat down, put the pen back on his desk, and outlined my question.

As I talked, I observed what *Doctor Clutterfree* was doing. Although he was listening, he seemed preoccupied somehow with the whiteboard pens that lay on his desk, and the one that I had returned in particular. I had not put it back exactly where it had been.

As he listened, he started to fiddle with the pens. He moved them about. He rearranged them once. He rearranged them twice. Not satisfied, he kept on rearranging them in several ways as I outlined my query. Finally, he stacked all six into a triangular array, perfect in its symmetry, at the front of his desk. Now satisfied with the perfect order on his desk, he responded to my query. I recall that when he glanced at his whiteboard, he seemed almost indignant about me even writing out equation (15.1), as if it was wrong, a mathematical heresy, and outside his universe of legitimate mathematics. Besides that, I got the impression he wasn't too happy about me putting marks on his otherwise pristine whiteboard.

I came away from that meeting with two new insights.

First, that *Doctor Clutterfree* was too conditioned to be of any use to me. He couldn't contemplate the validity of equation (15.1). He insisted that the rigorous *Riemann-Stieltjes theory of integration* required the *lower* limit of a one-dimensional integral to be numerically less than the *upper* limit, so that one side or the other of equation (15.1) was illegal.

Second, I realized from watching *Doctor Clutterfree* that his obsessive tidiness was probably a displacement activity. What I haven't mentioned up to this point is that, years before I even came to the department, *Doctor Clutterfree* had ceased to do any active mathematical research. Possibly, rearranging his pens ate too much into his research time.

My best advice here is to watch out for displacement activities. They serve no long term useful purpose and can seriously damage your productivity.

Paperbergs can be useful, if they are controlled. For instance, during a morning's work on some calculation, you do not want to be filing sheets of paper every five minutes. Put then in a nearby pile, ready for quick access. Sort out that pile at the end of the day, or it will grow and get out of hand.

You will often come across people with huge paperbergs on their desks and who seem to know where this or that paper is in those paperbergs. I have no doubt that their subconscious memory processes are keeping track of important papers. The problems arise only when paperbergs get too large for quick document retrieval.

Looking at the equal dangers of untidiness and over-tidiness regarding paperbergs, I think that mathematical physics becomes more like an art form than a strict science. Do organize your office and desk up to a point, in a way that works for you, but don't make it a religious observance.

## 15.2 Dating your work

This will probably sound a small point, but it's really important: *date your work.*

There's a very, very good reason for this. Suppose you have been doing loads of calculations for a couple of years. Your papers get scattered around your desk and office, and lie around in huge piles. Then one day, you're in the middle of an important calculation but you've forgotten a particular result you did last year. Where is it?

Ah yes, it's in that enormous paperberg in the corner. I'll dig it up.

At that point, you will experience the mathematical physics analogue of what paleontologists do for a living. They dig through layers and layers of old rocks looking for fossils. But, unlike you, they will usually have a good idea where to dig. They will have dated the rock layers, so they will know that they're in a *Permian* rock layer (dated 250-300 million years ago), rather than say the *Devonian* period (dated 360-420 million years ago).

Then there's *you*. You've done loads of calculations, but never dated your pages. You have *several hundred* loose A4 hand-written pages in no obvious order. You'll be lucky to retrieve that result.

Dating your work is an absolute time-saver, especially in the long run. I date even rough calculations, usually. I've often dug out old piles of A4 sheets several decades old and managed to reconstruct some ancient calculation, simply by looking at the dates in the top right hand corner of each page.

Dating my calculations is now part of my conditioning. I advise you make a serious effort in that respect. You may one day find that the calculation you've done today is still of interest to you, and you'll need to know when you did it.

## 15.3 Tagging and labelling

If you're involved in a project, there's every chance you'll be working on several separate items (calculations or notes) on the same day. Therefore, simply dating the pages of your work on a given day may still result in long term confusion. My solution is three-fold. First, every separate item is given an informative title. This helps in indexing, discussed below. Second, each separate item is randomly assigned a three letter tag, such as **ABC** or **MTE**, and this is written on the top left of each separate sheet of that item. This ensures pages from different items can be identified. Third, successive pages in each item are numbered. I can reassure you that many students do not take this elementary precaution when submitting their loose pages of coursework.

## 15.4 Indexing

After a few years, the numerous dated, tagged, and paginated items that you've created will have been stored in some archiving system. Then comes the day that you need to find some information about, say, the Schrödinger wave equation for the hydrogen atom. Where is it?

That's where an *indexing system* comes into operation. A good index will help you pinpoint the archive where your items are stored and then you can retrieve them quickly. After many years' trial-and-error, I eventually settled on a spread-sheet solution. Items are entered in an EXCEL spreadsheet, with their dates, tags, titles, and archive location entered in appropriate columns. In the case of the hydrogen atom I mentioned just above, I would search the spreadsheet with the key words *hydrogen* and *atom*, and the search result should normally tell me where to look in my archives.

## 15.5 Backup

One of the risks of working with computers is the sudden and irretrievable loss of data, due to system failure, hacking, or even theft. If you do not systematically and routinely backup up your files, be prepared to waste valuable thinking time in redoing many hours of calculations.

These days, there are no excuses for getting caught out. Memory in the form of external USB hard drives is cheap and reliable. I backup up all my working files once a week on four or more separate hard discs stored in separate locations. With multiple copies, I am relatively secure against individual copy failures. I also store sequentially, meaning I have copies dated over several weeks or months. On occasion, I have had to go back to a copy made several months ago to dig up some files that had been erased on more recent copies.

Recently I invested in a solid state disc large enough to clone my entire main drive. Even if my Personal Computer was stolen, I could rebuild it with no data loss whatsoever, using my cloned disc as the new main disk.

Although all of this sounds a little bit over the top, it will take only one incident of data loss to convince you that regular backup is necessary. *Do not rely on one computer or tablet only. Sooner or later, it will break down.*

## 15.6 Reference book

One of the delights of mathematical physics is that you start to accumulate masses of formulae, theorems, observations, equations, and other relevant bits of mathematical technology. These begin to give you great power. For example, I found that in Volume Five of Gel'fand and Shilov's monumental series *Generalised Functions*, there was an astonishingly useful table of Fourier transforms involving spacetime singular functions, of fundamental importance in relativistic quantum field theory. With that table, great swathes of mathematical physics calculations suddenly became doable. I simply had to have that table easily accessible. I could not go to our main University library and take out Volume Five every time I needed to Fourier transform some singular distribution.

You cannot expect yourself to keep such details in your mind. If you've read Chapter 11, *Einstein's Brain*, you will recall that Einstein didn't see the need to memorize the speed of sound. He knew that he could always look that sort of information up.

My solution was to create a special hardcover notebook in which I wrote out significant formulae, such as Gel'fand and Shilov's tables, Maxwell's equations, Einstein's equations for General Relativity, and many more useful bits of mathematics. I still have that notebook and it's still relevant to my work. It doesn't need power and I can look over it in a few seconds. It simply serves as an extra memory.

These days, I've copied its contents to an online system that I can access at any time in any place as long as I can get onto the internet. You are well-advised to start organizing such a reference system for yourself, if you want to go into mathematical physics.

I would not be so foolish as to claim that any of my methods are the only way to do things, but you will find by hard experience that unless you organize your work along similar lines, your life as a mathematical physicist will become dominated by clutter and inefficiency.

# Chapter 16

# Giving a Talk

*The answers we have found have only served to raise a whole set of new questions. In some ways we feel that we are as confused as ever, but we think we are confused on a higher level and about more important things.*

[Kelley (1951)]

Sooner or later, anyone passing through school or university will be asked to give a talk to a group of people. It could be to report on a project that they've been working on for months, or an unusual and difficult interest of theirs, such as listening to music or watching football[1].

Being asked for the first time to talk to a crowd can be a nerve-racking experience. In this chapter, I'll discuss some points about giving a talk.

There are three kinds of talk that sooner or later you will have to give as a working mathematical physicist.

### *Lectures*

A lecture is a talk in a series of scheduled/timetabled talks, with the series aimed at giving the audience some familiarity and expertise in some extensive topic, such as learning a new language or about a new theory. Usually every lecture in a given series is given by the same person.

Lectures are generally aimed at presenting established knowledge to beginners rather than announcing new discoveries.

### *Seminars*

A seminar is a talk in a series of scheduled talks, such as once a week. Each seminar is given by a different, invited speaker, who may come in from some other institution just on that one occasion. Seminars are generally about recent research done by the speaker. Conference talks can be thought of as seminars.

Seminars are generally aimed at introducing recently completed or still in-progress research to experienced professionals in a given field.

A few decades ago, a seminar would have been a live experience, with the speaker and the audience in the same room. I'll qualify such an event as *live action*. In a live action seminar, people can interrupt with questions or arguments as to why the speaker is wrong. Members of the audience can sometimes heckle the speaker, trying to get them to shut up. I've even seen a member of an audience stand up and harangue the speaker at length, accusing them of plagiarising their work. I've seen it all.

These days, we have the growing popularity of video talks, where a person gives a live seminar in one University, say, and is watched by potentially vast numbers of people in real time in other Universities, which could well be in other countries. There is also the possibility of watching a pre-recorded seminar over media such as *YouTube*. These modern alternatives have their uses, but as with seeing a play in theatre versus seeing a film in a cinema, the presence of speaker and audience in the same room gives the most memorable experience, usually. For that reason, mathematical physicists still travel all over the world to participate in live action conferences.

### *Talks*

A *Talk* in its loosest sense could be a presentation by a school student giving an informal account of their project to their class. Whether anyone in the audience learns anything is generally less important than the entertainment value of the talk and the experience gained by the speaker.

Talks can be part of a project examination, in which case they should be given as if they were seminars.

### Stop Talking

One of the problems with live action talks is that *you cannot always switch them off.*

There is an old *Tom and Jerry* cartoon that sums up what I'm driving at here. The action takes place on an ocean liner. Tom (the cat) is chasing Jerry (the mouse) towards the rear of the ship (the stern). Tom leaps at Jerry. Jerry ducks and Tom goes flying overhead. Tom gets carried by his momentum over the end of the ship. Tom hangs there for what seems like an eternity, staring at us with a pained look on his face. He knows he's going to fall into the sea. But as he starts to fall, he twists around to face the ship, and digs his claws into the steel sides. Then he sinks slowly down, with high-frequency screeching from his nails digging into the metal, all the way into the sea.

I've been to meetings where a speaker was just like Tom in that cartoon. They had run out of time but you wouldn't know it. They had been assigned perhaps a twenty minute slot, or even an hour, but they just wouldn't leave. Just like Tom, they would cling to the podium at all costs, and had to be dragged out of there, kicking and screaming by the conference organizers.

A good *Chair* (person who introduces the speaker and controls events) would be ready for such characters. I've seen a well prepared Chair hold up a large card with the logo "two minutes", two minutes before the required end of the talk. An even better-prepared Chair would hold up another large card with the logo "0" , right at the appointed end of the talk. On one occasion, I saw a Chair hold up a pre-prepared large card with "−5" (minus five minutes) for a speaker who had a reputation for running well over time, had done so on this occasion, and not going to leave without a fight.

Sometimes, when I was at a talk where the speaker kept droning on and on, well past their time, with no action from the Chair, I would start to daydream. It was always the same idea. What if every lecture room, every seminar room, every podium, was built over an automatically timed .... *trapdoor*, with rotating knives waiting ... That would solve the over-running speaker problem, without a doubt.

I give the above anecdote to illustrate one of the serious mistakes a speaker can make, that is, *running over time.* Don't do it.

This brings me to the question of *how* should you deliver your talk? You may work out an answer from the following real-life incident that happened to me.

## Project Talks

Working in a maths department gave me many opportunities to see great students doing great mathematics, but sometimes, project talks didn't come across so well. Here's what I witnessed one afternoon, a few years ago.

In those days, Final Year maths students were often given a project to work on over a semester, either alone or in groups. On this occasion, I was asked to assess three students, each of whom had worked on their own project under the supervision of a member of staff and now had to present a twenty minute talk about that work. Let's call those students *Scruffy Boy*, *Power Girl*, and *Bright Boy*.

First on was *Scruffy Boy*. I must say I was not impressed with his appearance. He had not dressed up for his talk. He had sloppy, dirty jeans, a creased T-shirt, and scruffy trainers. *OK*, I thought, *mustn't misjudge. Lets see what he's got to say.*

It was clear right from the start that *Scruffy Boy*'s talk was going to be bad. The first problem was his deportment. He stood and gestured to us as if he was in a public bar talking to his mates drinking pints of beer after a football match. He slouched. He mumbled to himself and didn't look much at the audience, as if he were afraid of us. He slapped a transparency over the projector but it had no name or title on it, just some hand-written equations that made no sense to me. He mumbled some vague words as he pointed to the screen, but I didn't catch any of them. There was nothing about his talk that was memorable, apart from his appearance and delivery. There were some marks on the screen that could have meant something, but I felt as if he wasn't really telling me anything about his subject. It was, in effect, deadly boring and being presented as such.

Next up was a young woman, *Power Girl*. Immediately I could see a difference. First of all, she had clearly dressed for the event. She wore nothing outrageous but a refined business-like outfit. She looked professional and well-poised. Her informative overheads started with a title, her name, and a description of her project. She spoke clearly and I followed what she had done. I knew that in a real sense of the phrase, she *outclassed Scruffy Boy* in every respect.

What I though was a good point was that each of the three students listened to the other two talks. So at least *Scruffy Boy* should have learnt something from *Power Girl*'s presentation.

Finally *Bright Boy* stood up. *Oh no*, I thought, *Not another Scruffy Boy! Bright Boy*, like *Scruffy Boy*, was in jeans and T-shirt.

How wrong I was. *Bright Boy* gave a really impressive performance. Despite the fact that he looked like *Scruffy Boy*, *Bright Boy*'s talk was excellent. His overheads were as good as *Power Girl*'s: informative and well structured. He spoke eloquently and I understood and appreciated his project.

You can guess how I marked the three projects. *Scruffy Boy*'s project got a pretty low grade. I gave a First Class mark to *Power Girl* and a First Class mark to *Bright Boy*.

There is a moral here. *Power Girl* showed that *presentation* cannot be underestimated in public speaking. *Bright Boy*'s talk showed that if you've got great material, then your audience will forgive your personal presentational deficiencies. After all, it's the material that counts, really. I was marking the project, not the student, after all.

But it does help if the audience can see that you've taken a bit of care in your presentation. Don't make the mistake of thinking that dressing properly for a presentation is caving in to establishment pressure to conform or that being scruffy is cool. It's not cool, it's being ignorant about basic psychology. Audiences really do like to believe that the speaker has taken a little bit of care to present themselves and their material.

Having said all that, it's well known that Einstein was a bit eccentric in his appearance. Apparently, he didn't like to wear socks and his hair looked as if he was holding a Van der Graaf generator. But who cares? His ideas were great. He was the original Bright Boy.

## 16.1 Perspective

Whether you give a talk, seminar, or hand in a project report, it's a wise policy to have some common sense about what you're doing. Mathematical physics is supposed to be about the real world, after all.

Here's an anecdote about having no perspective in project work, typical of beginners.

**To The Last Inch**

In the maths department where I worked, we occasionally changed the degree structure. At one time, First Year students were assigned group projects. Here's what happened when one group submitted a report on their group project and I had to mark it (I was not the supervisor).

Their project was on *jumping out of an aeroplane and landing by parachute as close to a designated target as possible.*

They had decided that they needed a theoretical model, which is inevitable in such cases. In their model, they assumed that the plane was flying uniformly *exactly* two miles above the ground, towards the target at *exactly* one hundred and fifty miles per hour. The parachutist jumped out of the plane *exactly* a quarter of a mile from the target. The acceleration due to gravity was assumed to be *exactly* thirty two feet per second per second and there was no wind relative to the ground. Air resistance was assumed to be dependent on the instantaneous speed of the jumper, with a constant of proportionality that was read precisely from an online source.

Blah, blah, blah. They wrote out some details of their ideas and how they arrived at their final result. As I read it, my initial thoughts were *Not so bad for beginners, really.*

Unfortunately, they floundered right at the end. They quoted their result, which was how far off target the jumper would have landed ... *to several significant digits accuracy of an inch.*

Can you see the problem? The moral here is, *have some perspective on your work.* It's silly to quote accuracy to thousands of an inch if the basic starting assumptions are in terms of miles.

Is this relevant to mathematical physics? I think it's extremely relevant. Mathematical physicists have to have a perspective on what they're supposed to be doing, which is *to understand the real world*, not hypothetical unphysical conjectured worlds that have no empirical content. Here's another anecdote involving a First Year project, but this time, a doctoral student that I was involved in assessing at the end of their first year.

### What Does It All Mean?

The postgrads that I came across in the Maths Department were invariably really smart, bright young people with mathematical ability and talent radiating from them in waves.

On this occasion, I had to listen and assess a brief talk from a brilliant First Year postgrad about the research (and this would have been real research in the true sense of the word) that they had done over the year with their supervisor. The subject was *Quantum Gravity.*

I can't emphasize this point enough. *Quantum Gravity* is a real *Sheldon Cooper* topic. It involves *very* sophisticated mathematics and high-flying assumptions about space, time and matter.

There's only one thing wrong with it. There's currently zero empirical evidence for it. None whatsoever. In fact, it's more like a religion. You either believe it or you don't. If you want my view on it and some other topics such as *String Theory*, read Chapter 6 (*Nullius in Verba*).

I sat there and listened to a marvellously technical discussion, touching on concepts such as manifolds, categories, functors, spin networks, and such like. As a piece of mathematics, I could not say much about it. It was clearly way beyond my personal experience.

If that surprises you, just think about it. It's inevitable that a postgrad working solely on one specialist topic over several years can quite quickly become *the* Departmental expert on that topic, more knowledgeable on the details than perhaps their supervisor (who will invariably have to deal with many other students and may not have had the chance to learn the details of the project that they themselves invented and assigned to the student).

So what did I do?

In such circumstances, I always ran for the one reliable perspective that never fails: *common sense.* The student was *supposed* to be talking about quantum mechanics, which is to do with observers and measurements. And also about *gravity*, which also has to do with observers and measurements. So I asked, at the end of the talk, *Where are the* ***observers*** *in your theory*?

**Answer came there none.**

The speaker seemed puzzled.

If anything, I got the impression from the mutterings of others in the audience (who were quantum gravity merchants) that I had asked a forbidden, if not stupid, question.

Judge for yourself. Was that talk about mathematical physics as I've defined it to be elsewhere in this book, or was it just mathematics that had no contact with reality?

## 16.2 Know your stuff

Giving a talk requires *preparation*. If you don't prepare, you may well look like a fool or worse. So what must you do before your talk?

The first rule, I'd say, is to *know your stuff*.

This means making sure you are not just repeating what you've read online or in a book. If you're explaining a theory, make sure you understand it. If there are calculations you're describing, make sure you could do them, if asked, without notes, at least in outline.

### You Know Nothing

There's little worse than being verbally abused in public, especially by someone higher up in the food chain of science.

This incident was at another one of those annual Christmas meetings of UK theorists I've mentioned before. I've given anecdotes elsewhere in this book about incidents at these meetings that were plainly funny. This incident still makes me cringe, even though I was in the audience (at the back as always) and had nothing to do with it.

Most of the hour-long talks at such a meeting are given by real experts. At the end of their talks, there is always five or ten minutes for questions from the audience. Those can lead to interesting and informative discussions.

On this particular occasion, a youthful post-doctoral fellow had been asked to review progress in String Theory over the previous year.

You may have got the feeling from elsewhere in this book that I've no time or sympathy for String Theory. I explain why in Chapter 22, *Strings and*

*Things.* But I'm not exactly alone. Quite a few theorists either have never thought String Theory was the be-all-and-end-all that its proponents claim, or else they once worked in it and have seen the error of their ways, such as Lee Smolin [Smolin (2006)]. Suffice it to say that String Theory does have its critics.

Unfortunately, on this occasion, there was a rather hostile member of the audience listening to this young postdoc "reviewing" the grand achievements of String Theory over the previous year. The speaker's misfortune was that this hostile member of the audience had recently won a share in the *Nobel Prize* for Physics.

It was clear as the young man's talk droned on that the subject was bogged down in a morass of technical details that had little or no obvious contact with the real world of empirical physics (meaning, the theory still couldn't explain any real data better than other theories). When the talk finished, it was question time.

The Nobelist was clearly not sympathetic to String Theory. He asked a hard question. The speaker hummed and hawed, but failed to answer directly. It was clear he was lost.

That's when the Nobelist struck. In a clear, loud voice, he shouted out for everyone to hear that the speaker *should not have been reviewing the subject if he didn't know anything about it in the first place* (or words to that effect).

I can think of few situations in science that would destroy a person's self-confidence with more certainty than that.

On the other hand, when I recall this incident, I think that String Theorists deserve to get all the flack they can attract.

The moral is clear: *know your stuff.*

## 16.3 Check your technology

Besides the inevitable critics, other dangers lurk behind the lecture hall curtain, waiting to mug the unwary speaker. Here's some real-life incidents that I witnessed.

## Astronomer Royal

To see how funny this story is, you have to know something about Prof Brück. He was an extremely distinguished astronomer born in Central Europe and he looked the part. He was in fact the *Astronomer Royal for Scotland* at that time, so that tells you just about everything. He often wore a polka-dot bow tie and in his elegant suit and glasses looked *exactly* like the image of a grand elderly scientist that you will have seen in many a film.

I mentioned in Chapter 9, *How It All Started for Me*, that I changed from Astrophysics to Mathematical Physics at the end of my first year at University. But that was not because I lost interest in Astrophysics. I kept up my membership of the University Astronomical Society. Eventually, in my final year, I found myself the seminar organizer for the Society. That meant finding people to come and give an hour's talk to the Society on their work in Astronomy.

On this particular occasion, the Society committee had asked me to invite Prof Brück and he had kindly agreed to come. So one night, there we were, in a Maths Department lecture room, waiting to hear the good Prof.

He sat there, waiting to be introduced. Our president, Harvey, stood up, said a few words of introduction, and then the good Prof got up and switched on the overhead projector. In those days there was no internet. No power point. The choice was chalk board or transparencies projected onto a large white screen, and the Prof had chosen to project.

The problem was, nothing happened. No light came on. The projector was dead. In panic, we the committee rushed to the podium and started tinkering with the projector. We pressed buttons off and on. We shook the thing. We pulled out the plugs and put them back in. We replaced the bulbs. We replaced the fuses. Nothing. Zilch. Nada. Zero. And all the while, the Prof stood there, watching us idiots getting nowhere. Then, to my amazement, he acted. He asked us to stand aside, which we did.

Then, without hesitation, this bespectacled, bow-tied, highly reputable, mild-mannered senior academic gave a mighty **kick** to the mains electricity plug, which was at floor level.

And the lights came on and the projector worked perfectly after that.

I think the following has happened to other people too.

## Wingdings

I once had the great pleasure of going to Orlando in Florida for an international conference on *optics*. Believe it or not, that's a great field for mathematical physicists to work in, because both classical optics and quantum optics are fabulous disciplines to work in. Classical optics is used to design the lenses and mirrors that are used so much in modern technology whilst quantum optics allows physicists a relatively inexpensive way to explore the strange world of quantum mechanics.

Did I say conference? It was more like a mass migration. Normally, the scientific conferences I've been to attracted at most a few hundred delegates. *This* particular one in Orlando had *five thousand* delegates.

I absolutely loved the experience. First of all, I really like the Americans at home. Whenever I visited their towns, stayed at their hotels, or dropped into their homes, my experiences have been excellent.

The conference talks extended all week, from Monday to Friday. With so many delegates, there were multiple talks going on at the same time. My particular talk was relatively early on in the week and it went reasonably well. There had been a small audience, but that was typical for the branch of optics my talk had been slotted into. I gave my talk and settled down to visiting any of the other talks that took my fancy.

I was due to fly back home on Saturday, so after visiting Cape Canaveral Space Centre on the Thursday, I listened all day Friday to the very last talks of the Conference.

There is something sad and moving about conferences as they reach their last day and last talks. Some of the friends you have made in the early days of the conference may have already gone home by the last day, and you may perhaps never meet them again. If you are relatively low on the ranking order (as determined by factors such as status), then it's possible you will be close to the last, or even be the last, speaker.

There I was, sitting in a poorly attended lecture room, with the very last speaker about to start his talk. I had seen him throughout the week, so he had waited five nervous days for his moment of glory.

The young man was introduced by the Chair, stood up, connected his laptop, and started his half hour talk.

The only problem was, the fonts that were displayed were all wrong. By some strange twist of fate, his laptop was displaying all his equations in *Wingdings font*. For example, instead of an equation such as

$$ax^2 + bx + c = 0,$$

all we saw on the screen was

♋⊠⌹☑♌⊠☑♍⧫□.

There was clearly something going wrong with his fonts, and no one could fix it. That was the end of his talk. All that the Chair could do was to advise the speaker to email a pdf of his talk to anyone in the audience who was interested.

The sad thing is that the speaker could have checked his technology any time during the week. He didn't and his talk went down the drain in consequence.

I don't know about the next anecdote. I don't think there's much anyone could have done in the given circumstances.

### The Universe is out to get you

*Conferences* and *Summer Schools* are meetings where theorists give talks on their work and listen to what other people have been up to. A good venue is a bonus. You get to hear great talks on fascinating subjects, in exotic places you would never have visited otherwise.

This anecdote is what happened during a Summer School in the City of Durham. It's a lovely, ancient city, with an impressive history, architecture, University, and scenery.

On this occasion, it was just after coffee time in the morning. The last talk before lunch was to be given by a young researcher in some branch of elementary particle physics. I'm not sure what the talk was about, because no one got to hear or see it.

The talks were all held in the biggest lecture theatre I think I've ever seen. It was *vast*. It had hundreds of seats and it was tall. The roof was curved and seemed to be a couple of miles high in the sky. If you've ever been to the *Pantheon* in Rome you may have some idea what I'm talking about.

The throng of experimentalists and theorists piled into the lecture theatre after coffee, but there was space aplenty. The place was so big that I didn't sit at the back as I normally do (I had no opera glasses). Even though I sat at the back of the *audience*, I was in the middle of this Colosseum-like lecture theatre.

The morning session Chair introduced the speaker, who stepped up and switched on his notebook, from which his Powerpoint slides should have been delivered.

Nothing happened. Somehow, the connection between his machine and the lecture theatre's projectors would not work.

The organizers eagerly stepped up and adjusted the equipment. They pressed buttons, fiddled around with leads, and exchanged projectors, to no avail.

In desperation, one organizer suggested transferring the speaker's Powerpoint presentation from his notebook onto another notebook that was known to work, and hopefully, the program would run from that one. Five minutes later, a suitably prepared USB stick was pressed into the new notebook and everything started ... to go wrong again. For some reason, no image came on. Some problem in the software or hardware meant that we were not going to see any of the speaker's talk on screen that morning.

There were agonized moments of despair and frustration, until the ever-resourceful Chair had a great idea. Behind the speaker there was an enormous whiteboard. Could the speaker give their talk and draw relevant diagrams and mathematics on that?

The speaker gratefully agreed. It would be a reasonable fall-back plan.

The speaker picked up a whiteboard pen, turned to us and opened his mouth to start his talk.

That's when the Universe intervened. At that moment precisely, the gods of meteorology decided that a thunderstorm of torrential proportions would start to play a full-volume quadrophonic version of a Karlheinz Stockhausen symphony on the acoustically unbaffled roof of our cavernous lecture hall.

Have you ever been near a loud waterfall? The roar of water can be quite deafening.

We all sat there, watching the speaker open and close his mouth as he started his talk. All we heard was a sound of a thousand waterfalls reverberating in the echo chamber that was indeed that lecture theatre.

I think the unfortunate victim gave up after about five minutes. It was clear the Universe was not going to let him give his talk that morning.

The moral here is to check your technology well *before* you are due to give your talk. Have a backup plan, prepared well in advance, in case technology goes wrong.

## 16.4 Be aware

The following anecdote is about a talk a fellow student gave, many years ago in school, about his hobby, *photography*.

**Watch what you say**

There we were, several hundred teenagers, sitting in the School Hall, waiting to hear *Bright Spark* give a talk on his hobby, which was *photography*.

There's nothing wrong with photography. Both the picture taking side of it and the processing side of it are quite fun. It's just unfortunate that *Bright Spark* did not quite realize the effect his choice of words would have on us when he introduced his subject with the immortal words...

"*I first became interested in photography when my brother started developing...*"

I can't quite remember what the rest of *Bright Spark*'s talk was about, as we didn't stop laughing for several weeks after that.

The moral here is to have an awareness of how what you're actually saying comes across to an audience.

## 16.5 You are what they see

There's an important point I should make about giving a talk. It's about *impression.*

What I mean is this. Whenever you stand up before a small, large, or vast audience, you're setting yourself up to be *examined.* Not by writing answers in an examination book but by your appearance, by the way you speak, by the way you stand, by the way you interact with the audience, and of course, by the content of your talk.

It's a hard fact of life that us humans can be a pretty cynical bunch when something comes across to us in perhaps an unintended way. An audience *could* get it into its head that *you* the speaker are wasting their time. If that happens, you may have an impossible job overcoming their complaints. *Stand-up comedians* will tell you of the numerous times their act had died once an audience had starting its background rumblings of discontent.

What you're dealing with here is the psychology of an audience, or in other words, whatever thinking patterns they have been conditioned into. How do you fight that?

I'd say that there's just one approach that gets you out of trouble every time: *be professional.* What I mean by that is this.

First, *never* be frightened of a crowd. I remember that when I first started to lecture to classes of a hundred or more students, I was pretty nervous. That is, until I suddenly saw the truth of the matter: *as far as any individual member of an audience is concerned, there's only one of them.* So when you talk to an audience, you should think that you are talking just to one person. I found that particular piece of self-conditioning settled my nerves on many an occasion. That is, whenever I remembered it.

What happens if this strategy goes wrong? What if the audience is starting to talk and the background noise, the *audience buzz*, is slowly rising, threatening to drown out your talk?

There are several ways that I've seen speakers deal with this. Some speakers ignore audience buzz, but that's unfair on those members of an audience who want to hear the speaker, not the people around them. So don't ignore audience buzz.

An inadvisable response to audience buzz is to confront it head-on and tell members of the audience to shut up. There are two dangers here. First, if it doesn't work your authority will be clearly threatened and audience buzz will probably get worse. Some students actually take delight in seeing a teacher or lecturer show that they have lost their cool. The second danger is that you might alienate people in the audience who were actually listening to you. So I don't advise confrontation.

A response to avoid at all costs is to walk out. There are two problems with that strategy. First, it looks unprofessional, even if you believed that you had no choice but to leave. I did that once or twice early on in my career, but there always came the feeling afterwards that I had lost the argument. The second problem is that unless you're a Nobel Prize winner that everyone really really wants to hear, a student audience will generally love you to walk out. They'll even open the door for you, because that means they now have some more free time.

Eventually, the method that I found works really well is to **get the audience on your side**. A talk is not meant to be a confrontation. A talk should be a collaborative effort by a speaker and an audience to travel together towards new understanding. It's not a case of *speaker versus audience*, but *speaker guiding audience*. I mean, if you were guiding a party of tourists up *Mount Everest*, your trip would be a disaster if you and the tourists were at loggerheads.

An effective way of getting an audience on your side is *humour*. For instance, whenever audience buzz started in one of my lectures, I would ask people on the right-hand side of the room to *Speak up please! The guys over there on the left can't hear what you're saying*. What invariably happened then was that initially, there would be a lot of laughs from all around and the audience buzz would actually go *up*. It would last only very briefly, and then it would usually go to near zero. I had made my point, non-aggressively.

This next anecdote illustrates that a speaker should be a bit careful *how* they speak.

### Um, you know

I was fortunate to have worked for most of my career in a maths department. I came across many fine students, talented and excellent in diverse ways. Of course, there's always room for improvement in anyone, as they say.

On this occasion, one of my personal tutees, *Clever Boy*, came to see me. As you will know by now, a personal tutee is a student that you, the personal tutor, deals with on a one-to-one basis. You welcome them into the department when they first arrive and you guide them through the bureaucracy of student life over a year or more (such as Career guidance).

*Clever Boy* was an exceptionally well-organized and meticulous student. He was the only student I ever came across who would actually prepare *two* copies of his end-of-year personal statement when he came to see me, one of which was for me and the other for his own records. I don't recall any other personal tutee of mine ever giving me a single previously prepared personal statement, I should add. In short, our departmental scheme for administering students must have been modelled on *Clever Boy*. He was the perfect role model in that respect.

In addition to that, *Clever Boy* was an excellent, bright student. He invariably passed his exams well and caused me no problems whatsoever over the three years involved. When in his final year he came to see me, it came as no surprise to me that he was applying to a prestigious University in London for a higher degree. I was confident he would get in, and I told him so.

Far from being pleased, he was concerned. He was a natural worrier. He told me that he was to be interviewed in London in a couple of weeks and would have to give a ten minute talk on the Final Year project he was working on currently in our department. Therefore, he had prepared his talk and would I help? He wanted to practice his talk in front of me, so I could give feedback.

I was very pleased and impressed with *Clever Boy*'s approach, and so I agreed. I went online to the University's booking system and found that a large lecture theatre was available at a certain time in a day or so. We agreed to meet there and then for his practice talk.

Came the day, came the hour, there we were. I sat right at the back as I always do and he stood at the podium with his talk ready. Being at the back, I could check whether he spoke clearly enough to be heard. I gave him some standard pieces of advice, such as not to move around too much and not to over-run the scheduled time, and so on. He then began his talk and I started my stop-watch.

Now his subject matter was good, I could not fault him on that. But the more he got into his theme, the more I realized there was a clear problem. He was *ummming* frequently, and inserting *you knows* just about every other word.

I began to count them. I had brought a sheet of paper with me in order to take notes on his talk and then give him feedback. I started to put a stroke on the paper for every *ummmm* and another stroke for every *you know*, a bit like you see in films where someone is in a prison and doesn't have a calendar.

When he finished, I counted them up. There were over seventy *ummms* and over a hundred *you knows.* I'm not exaggerating.

When I showed him my notes, he was crestfallen, until I reminded him that the whole point of this exercise was for him to find out what problems there might be with his talk. His material itself was fine, but the delivery could have been much better.

Knowing *Clever Boy*, I'm sure he gave his talk with improved delivery in the actual interview.

## 16.6 Whatch you're spelin

In addition to the impression created by your delivery in a talk, there is another factor that any speaker or author should take into account. That is *spellin mistake's.* If there's one thing that puts me right off any piece of writing it's in spelling mistakes.

You might well disagree. After all, you could say that language is constantly changing. Language is what ordinary people speak, not what stuffy professors of English (or any other language for that matter) have decided.

Here's what I have to say on this matter.

First, careless use of language in the written word is essentially *permanent.* I mean, what you have written may be read as it stands many years later. I recently read about the famous *Book of Kells*, a fabulously illustrated manuscript book dating from around 800 C.E. There are a number of differences in that book between what it says in places and what is written in authentic versions of the Christian Gospels. It is clear that the Book of

Kells contains errors. When *you* write, whether for an overhead in a talk or for an article in a journal, keep in mind that your spelling may haunt you for years.

Second, spelling and grammatical errors reveal your mind and how you've been brought up. For instance, if you swear a lot in normal conversation, you will soon find that swearing during a public talk will damage your reputation. For one thing, such language will suggest that the material you are talking about is weak and requires artificial verbal boosting. It's not advised. Good science speaks for itself and does not need hyping up.

A third point here is that standard use of language makes it easier for non-native language speakers to understand. In this respect, I am frequently impressed by the high standard of English from researchers whose first language is not English. Whenever I read a well-written paper by someone who had to learn English as a second language, I tend to think they know what they're writing about. It may be unfair, but a paper with poor grammar usually fails to impress me.

This last point deserves further commentary. Bad grammar is one type of poor writing. Another is the use of expressions that come and go out of fashion like the wind. I remember reading an email from a personal tutee who wanted to be advised about doing a higher degree in Astrophysics after they had graduated. "Cosmology is so sexy!" he wrote.

At that time, *sexy* was used frequently to mean *cool*, a word that I think still means *greatly respected* to the person in the street. As I read that email, I winced. I never though cosmology had anything to do with reproduction. Of course, I did not have the bad manners to correct that student.

## 16.7 Sales pitch

In principle, you give lectures to give technical details. You can get a lot of detail across in ten or twenty lectures. In contrast, the limited time given for a seminar or talk doesn't allow for much detail to be given. So what's the point of a seminar or talk?

The aim of one-off talks is to *stimulate* an audience into getting more interested in your subject than they were before you started. It's not important whether or not they actually learn something new in detail. That can come afterwards, if you've stimulated their interest. After all, they have the rest of their lives to find out more of what you been talking about, but only if you've given them that motivation.

It's easy to overlook this fact and overdo it. Looking back, I think that I overdid things in several of my conference talks. I would usually prepare twenty or more overheads (equivalent to pages) for a twenty minute talk. Each page would have masses of words and equations. I usually forgot the fact that reading a page of text takes a person much more than a minute to grasp fully, especially if it's a technical subject. Reading should be treated as a fine meal, consumed slowly to appreciate its delights. Speed-reading is all very well, but can you remember anything afterwards?

It's a mistake you see quite often. A person may have worked on a highly technical theory for a couple of years. Then they are invited to give a talk at a conference. They go there, they are introduced, they switch on their overheads, and the audience is then faced with masses of text and symbols that they have no time to absorb.

## 16.8 Don't avoid talks

I came across many students in my time. Some of them were reluctant to give a presentation, or even answer a question on my whiteboard during tutorials in my office.

I can understand that reluctance. The imagination can play havoc with our confidence. What if we make a mistake in public? What if we are criticized in public? Will we look stupid?

I would advise anyone, regardless of what they want to be, to never hesitate giving a presentation or talk. As a beginner, you will make mistakes. Rest assured, however, that provided there's reliable and fair feedback from the audience, you will get better and better the more you practice. Some people are natural speakers, but others have to learn how to do it properly. Remember that, if you do decide to become a mathematical physicist, communicating your ideas is an essential part of that career.

# Chapter 17

# Finding a Job

*If women are expected to do the same work as men, we must teach them the same things.*

Plato, The Republic

Sooner or later, you will go *job seeking*. There are three important reasons for getting a job.

## 17.1 Income

Normally, you will need to support yourself once you leave College or University. *That means getting a job.*

If by some chance you have parents who support you financially whilst you do nothing but go online and play computer games, then stop that if you can. I would advise you to re-evaluate your life and get out of that way of life. Of course, if you are unable to work because of some disability, being supported by others is in order and I would not quibble with that. But being severely disadvantaged by his medical condition did not deter Stephen Hawking from being a very active and successful scientist.

### Now That's Really Smart

As Departmental Careers Officer, I remember once sending out a questionnaire to our graduating students, asking for details about their prospective careers.

You can imagine my feelings when one third year student returned his form with the information that his *starting* salary in a Merchant Bank (also known as an Investment Bank) was to be *sixty thousand pounds* per year. At that time my salary as an established academic with a doctorate was *thirty thousand pounds* per year.

I do hope he didn't drown in all his money.

## 17.2 Self-development

Getting a job normally involves interacting with other human beings, which is generally thought of as a *good thing*. "*No man is an island*" is a useful saying that encapsulates this essential truth. When you get a job, you learn your limitations and your strengths. As a mathematical physicist, you will meet much smarter people than yourself. Even *Sheldon Cooper* had his *Dennis Kim* in *The Big Bang Theory*. You can learn from such people and become a better mathematical physicist.

I once had the privilege of meeting one of the great mathematical physicists of the Twentieth Century, *Julian Schwinger*. When he died, many of his former students and colleagues wrote tributes to him, describing how he worked. Unlike another iconic mathematical physicist of the Twentieth Century, *Richard Feynman*, who developed the diagrammatical notation known as *Feynman diagrams*, Schwinger developed his quantum field theories using sophisticated symbolic notation. Schwinger was a great master of the subject. He devised subtle and effective mathematical techniques for squeezing empirically testable predictions from seemingly impossible-to solve-theories such as *quantum electrodynamics*. Like Feynman's diagrams, Schwinger's techniques are still used in many areas of mathematical physics.

In those tributes to him, I read that Schwinger generally worked very much on his own as a theorist. He was in many senses of the term, the lone-genius mathematical physicist that I've alluded to elsewhere in this book. Nevertheless, when some new member of staff arrived at Schwinger's department, Schwinger would sooner or later drop in on that person and talk to them, to find out what they did, what they knew, and what he, Schwinger, could learn from them.

That might sound as if he was "stealing" other people's ideas. No, far from it. There's an important historical precedent for what Schwinger would do, and that's *Aristotle*, the polymath (universal genius) of Ancient Greek times. Aristotle was famous for collecting from other people as much knowledge about the Universe as he could, just like Schwinger. Rather than think he knew it all (he almost did), Aristotle knew that much knowledge and wisdom resided in the collective around him, and that it was a sensible strategy to be aware of it.

Rather than "stealing" other people's ideas, what Schwinger did showed intellectual honesty. The bad theorist thinks they know it all. The good theorist knows that they don't know it all and can learn from others. My point here is that one reason for having a job is to learn new skills and gain experience from other people. You can't do that home alone.

## 17.3 Treading water

This may come across as a strange reason for getting a job, so I'll explain it. As you get educated, in school, college, and possibly University, you'll acquire a great deal of knowledge about many technical subjects. But what you might not acquire is *inspiration*, *purpose*, and *hands on working experience*. There's every chance that a person finishing their first degree still doesn't know at that stage what they want to do next. To give themselves time to think, many people go around the world with their friends just after graduation. Others take any old job that comes along, giving them some time to get something better. That's what I mean by *treading water*.

As a personal tutor for many years to mathematics and mathematical physics students, I encountered a wide spectrum of talent and purpose in my students. There have been students with outstanding mathematical abilities and zero interest in any particular career, and students with terrible exam results who were desperate to become mathematical physicists. Given a choice, I would choose to be the latter, the person who has failed exams but has a clear focus on some career. Such a person might well pass their exams next time, and their ambition will drive them to success. Having talent and no ambition is by far the worse option. That's why, if you are in that position, you really do need to think about yourself and start to develop a sensible meaning to your life.

When I first started in the maths department, I was surprised to see how many students started their University life with no clear life ambition. I think I was fortunate in encountering teachers at school who inspired me by the time I was fourteen to become a scientist.

Over the years, however, I realized that not having any idea what to do after University was not unusual. Not everybody comes to the same place at the same time in life. It's such an important decision, *what to do for the next forty years*, that thinking about it for perhaps a year after graduation is reasonable. That's what I mean by *treading water.* My view now is that if you haven't reached a point in your life where you know what sort of career you want, *that's all right.*

However, young minds, like fine cheeses, do mature. If a person treads water for too long, they can lose focus and interest in preparing for life after graduation. As far as mathematical physics is concerned, most people who go into that career generally know what they're interested in by the time they come to complete their first degree, and usually they start on a doctorate (PhD) right after that. It's not necessary, however, to jump into your doctorate immediately. A year out in between graduation and starting PhD can be beneficial, but much longer than a year risks the potential PhD never starting.

Going to University while you think things out is a good strategy if you don't yet know what you would like to do in life when you leave school. In particular, doing a Mathematics degree, of any sort, is an excellent way to sort yourself out. You can sharpen up your mental faculties, you will be studying a subject of tremendous antiquity that will always be of value to you, and it can serve as a springboard into all sorts of careers.

One of my academic duties at one time was *Departmental Careers Officer*, arranging Careers talks and collecting employment statistics. I never thought it was any of my business to advise the students about going into any particular career. That was for them to look into and decide on. My job was to facilitate that process, by bringing in speakers from industry and science to outline various career options. I once counted all the professions that our mathematics students had entered over a twenty year period. I listed over *one hundred and ninety* different careers that our students had gone into, after graduating in Mathematics. Many of them were mathematically based careers, such as finance, accountancy, actuarial, software engineering (such as writing computer games), and teaching. A fair number went on to do a second degree. There were also some really exotic careers. Over twenty years, one of our students had gone into the Anglican Ministry, one or two had gone into the Radio and Television industries as executives, and so on.

I did not have any information, of course, about how many had been recruited into GCHQ (Government Communications Headquarters), the UK centre for security and intelligence. I suppose if I ever got such information, the student concerned would have to kill me.

I think the best piece of advice I gave to any student who came to see me about what sort of career they might go into was this:

"*Go for a long walk on your own around the University lake and imagine yourself in five or more years' time, waking up to start a new day. Do you want to drive for an hour to an office where you will meet up with a band of colleagues working on some long term project, or will you walk into a field where you are conducting research into soil management, out in the fresh air all day?*"

My point is, try to see in you mind's eye what sort of daily existence you would like. Some people want the well-ordered life of a company office, whilst others want to work in the open air. Some people want team work, whilst others want to work alone. Imagine you are going to have that life style for ten, twenty, or more years. Only you can decide what sort of life may suit you.

Finally, on this theme, there's one point to remember. Nothing is inevitable, nothing is forever. You are allowed to make career choice mistakes in life, but it's best to try to avoid them in the first place by a bit of thought and planning.

## 17.4 Finding job opportunities

Now it's the place in this chapter to change the theme. Suppose you have settled in your mind the sort of job you would like. There are several stages of the process that you will have to deal with. It's all very well me advising people how to apply for jobs, but how do you find out about them in the first place?

In my particular case, I solved this problem using what I call my *machine gun approach*. It really works.

### My Machine Gun

There I was, at the start of the last year of a two year postdoctoral research job, and the obvious questions had reared their ugly heads once again: *What and where was my next job going to be?*

At the time I was sharing a house with two other people in the same position. They were older and more experienced postdocs than I was and I imagine that given an application from all of us for the same job, one of them would get it. Don't get me wrong. There was no sense of competition between us in that house. We didn't discuss jobs. Most of the time, our evening discussions were about the Muppet Show and Monty Python.

As my second year in that post started, I realised that I had to have a definite strategy in order to get my next job. I settled on the following.

First, I decided I was going to apply only for jobs that my training as a mathematical physicist might come in handy. By that, I mean I would be *flexible*. Mathematical physics gives you that possibility. I would not insist on getting a job that matched exactly with my personal experience and ambitions. I decided I could endure a few more years doing any reasonable sort of science, if it involved some theory.

Next, I ordered certain newspapers and publications to be delivered on certain days to the house. For example, one of them was the weekly *Times Higher Education Supplement*. Those newspapers and publications carried numerous scientific and academic job adverts.

Once I received those publications, I then looked over them carefully, cutting out all adverts for any jobs that might possibly be suitable for me. I then pasted each cut-out advert onto a separate sheet of paper. Then I got to work, writing out a separate job application for every advert, and sending it off immediately. There was no email or internet then, so it was quite a chore and quite expensive in terms of postage to do this.

For every application I sent out, I recorded the date I sent it on the relevant sheet (the one with the pasted-on cut-out), and then ... *immediately forgot about it.*

That was the crucial aspect of my "*machine gun*". I had realized that being mentally attached to any job application is a great mistake. If you keep thinking about that one application for the perfect job, you will end up going crazy.

There's a good reason for this aspect of my "machine gun". Contrary to what you might believe, *there is not a conspiracy out there to deny you a job* (or so I like to think). No one is that important. Even if there was some prejudice because of say your name, colour, or gender, it would not be universal. I know it used to happen to ethnic minorities and women a great deal then, but there were always some employers who were not biased. Despite having an awkward, long, and non-Anglo-Saxon surname, I always believed that a job rejection had to be regarded as an impersonal event. I expected to get rejected, not because of my background, but because I really was unsuited for that job. The plain fact is, everybody gets rejected in some way, sooner or later.

I should mention here another reason why I admire Americans. They're so polite, even when they reject you. Some applications I sent to UK universities were never acknowledged or replied to. With the good old USA, *every* application I sent off came back with an excellent and encouraging reply, typically of the form: "*Dear George, Thanks so much for your interesting application. At this time, we have limited funds for further employment opportunities, but will gladly consider you should the occasion arise.*"

I'm not exaggerating. The first time I tried my machine gun approach, after I got my doctorate, I collected exactly forty really polite rejections from the USA. I kept them in a bundle for years as a souvenir. The fact that I never heard from any of them ever again was irrelevant. I felt encouraged by their positive sounding replies. At least they acknowledged my enquiries, which is better than having them thrown into a black hole.

The good news is that, even if you have been rejected hundreds of times, there will come, sooner or later, an offer of an interview.

Given that, you should treat job applications impersonally. That's where my "machine gun" analogy comes in. When you fire such a weapon, you saturate the target area with many bullets. There are many "bulls-eyes" in the target area. Those are the jobs you're applying for. Sooner or later, one of of your bullets will hit one of the bulls-eyes. Likewise, I sent out many job enquiries about many jobs, confident that sooner or later, one of them would be successful.

There is a rider to this anecdote. I got another two-year appointment using my machine gun. I kept on using that technique and finally landed a permanent job that lasted until I retired. After I started that final job, I returned to the first University for a brief visit and met my old friends back in the house we had shared three years earlier. They were still there, having renewed their temporary contracts with the same University. One of them asked me how I had found the advert for that final permanent job that I was now doing. He had been quite unaware of the advert. If he had seen it, it's more than likely that he would have applied and got the job. It's clear he did not have a machine gun.

I give the above anecdote to show that if you want to achieve an ambition such as becoming a mathematical physicist, *you must be organized.* Don't rely on blind chance. Be prepared, train hard, and take your opportunities when they arise. It's actually hard work getting a job, sometimes.

The following anecdote illustrates that sometimes, job opportunities may come to you in the most unusual of ways. If you don't recognize them when they come and seize the opportunity there and then, they may never come again.

### Carpe Diem

It was once again the annual Christmas meeting of UK Theorists, in the same place where Stephen Hawking's lecture had been interrupted by that security guard (see the anecdote in Chapter 2).

This particular year, there was tremendous excitement. A renowned cosmologist was going to start the show with a review of the latest experimental data and analysis of observations from the most distant galaxies observable. The data suggested that at the remotest observable distance scales, *space was flat and galaxies were accelerating away from us.* Scientists are still trying to understand the implications of that.

Unlike the time when I was applying for a postgraduate degree (a doctorate), when cosmology and particle physics were considered to be distinct subjects, the two areas of mathematical physics now overlapped significantly. There was now more and more evidence for a strange form of matter and energy known as *dark matter* and *dark energy* permeating the universe. To this day, no one knows for sure what this means. Some people think it consists of strange forms of particle, others think it is a manifestation of unusual gravitational force. I would like to think that someone reading this book right now might be inspired to become a theorist and find out. It's in your hands. My generation of mathematical physicists has no real ideas.

Getting back to the story, just imagine the excitement. Hundreds and hundred of academics, researchers, and students, from all around the UK and far beyond, all milled around outside the lecture theatre, waiting for the doors to open.

Then, on time, the doors opened as if by magic, and we all flooded in. And I mean, *flooded in*. I was in the crowd and bounded up the steps, right to the back row, where I usually sat. Quite quickly, the place was full. People sat in the seats and on the steps, and there was a great buzz of excitement as the main speaker was introduced.

The audience buzz gently abated and the cosmologist began. He switched on the overheads and started to give us a thrilling overview of the data, the equipment, and the great observations that had led to this fascinating new view of what our Universe looks like right at the limits of current observation.

He went on, for about fifteen minutes.

And then it happened ...

You will know by now that I always sit at the back in lecture theatres. What's at the front depends on the type of theatre. Sometimes, there's just a solid row of seats and desks, whilst in some theatres, the front seats have no desks. That makes it easier for "big-wig" scientific leaders and conference organizers, who invariably sit in the front, to stand up and get to the speaker's podium. So it was in this particular lecture theatre.

Someone sitting in that front row suddenly stood up, right in the middle of the speaker's talk.

The speaker stopped and stared at the man. The audience hushed and stared at the interloper. What was going on?

The man looked around self-consciously at the assembled horde of professors, post doctoral fellows, research students, all staring at him in baffled silence. Seconds passed but it felts like aeons. Finally, motivated by some call to duty, he suddenly blurted out ...

"*I've got to go ... I only came to make a delivery ...*"

And with that, he made his way for the exit and was gone, never to be seen again.

It was obvious what had happened. The man had come to make a parcel delivery and had found himself in the crowd, outside the theatre, before the doors were opened. When they were opened, he was dragged in with the flood of people and had found himself sitting in that front row, listening to the latest news about the structure of the universe.

He had, it seems, found it very interesting, if not fascinating, but the call of his delivery job had finally told him he had to go.

I occasionally think of that man and imagine this. Suppose he had decided, from what he was listening to, that doing cosmology was better than delivering parcels. What if he had decided to quit his job and ask the speaker about becoming a cosmologist? You may think that's fanciful, but something like that happened in the case of Michael Faraday.

One of the interests I have is reading the biographies of great scientists. I read about Michael Faraday. He was one of the great experimentalists of the nineteenth century, but he started life as an apprentice to a bookbinder. He was so influenced by reading the science books he was binding that he decided to become a scientist, and the rest is history, as they say.

One of the great ideas Faraday developed and championed was the concept of *magnetic and electric lines of force*. That has become a central concept in mathematical physics. It demonstrates the power of intuition, for Faraday was famously anti-mathematical and used visual displays to illustrate what he thought was happening in space around magnets and electric charges. I show such diagrams in Chapter 22, *Strings and Things*, which I use to explain the origins of *String Theory*. Do read about Faraday. You will be astonished.

I can imagine that if it had been Michael Faraday sitting in the lecture theatre that day, he would not have left to make his deliveries but would have stayed on to ask questions of the cosmologist speaker. He might have joined his research group as a tea-boy, and gradually worked his way up to a Nobel Prize in cosmology. Faraday did something just like that by starting to work as an assistant to the famous chemist Humphry Davy. At one point, Faraday even served as Davy's valet. As the Romans used to say, *Carpe Diem* (seize the day/moment).

Sometimes, we should jump into the unknown. Take a chance. Do the seemingly impossible. You want to be Einstein? No, it's not possible, but you can try to emulate his success. If and when the chance comes, don't hesitate. That opportunity may never come again. Do make sure you really want to do it, though.

## 17.5 Great bosses

Now I've gone on a bit about deciding on a job. But there's one crucial element in the job-hunting game[1], and that's the head of group or employer that's advertising the job.

Here's the real deal about mathematical physics. It makes you really versatile, but not everybody may know that. On a number of occasions, I used my "machine gun" to apply for jobs for which I had no experience. One of them was doing experimental physics (nuclear magnetic resonance) in the physics department at the University of Kent in Canterbury, another was doing computer simulation of liquids (water) in the physical chemistry department of the University of Oxford, and the final one was as a theoretical quantum chemist (quantum chemistry) in the mathematics department of the University of Nottingham. In all three cases, I had zero experience in those fields. How did I get the job each time?

I imagine more suitable candidates had either not known about the job (no machine gun) or else had been offered it and had declined it (it does happen). That still leaves the question unanswered: *I was not experienced in those fields, yet I got the job each time.* Why?

The answer is: *the head of each research group looked on me as a versatile person. Given my background in mathematical physics, they had confidence that, given a few months, I could learn my way into the work they were involved in.* I hope that I did not disappoint them too much.

Two job interviews I had did not end up with a job offer. In one case, I was told that the time I would need to embed into the project was too long given the duration of the contract, which is a fair enough point. In the other case, it was clear that I was overqualified for what was essentially a routine job that could be done by lower paid workers. Overall, my experience with the jobs I did get is that mathematical physics certainly gave me adaptability and I was able to fit into the jobs I got relatively easily. If you have been interested by my sales pitch for mathematical physics but remain worried by long term job prospects, I can say from personal experience that mathematical physics has opened up more career opportunities than it closed.

The same considerations about adaptability applies to students doing a straight mathematics degree rather than a mathematical physics degree. In both cases, the majority of graduates have no trouble finding employment in a huge range of careers.

# Chapter 18

# Interviews

*Choose a job you love, and you will never have to work a day in your life.*

Confucius

Interviews are one of those experiences that we all have to undergo in order to get accepted for jobs, degree courses, and social clubs. In the following, I'll talk about interviews as if they were all for *jobs*.

In my time I have had quite a few interviews. Most of them went well, but one in particular did not. I'll explain why in one of my anecdotes (*Tax Man*). As it turned out, I'm *really* glad I didn't get through that interview. I would like to share my experiences in some interviews, as this may help you if you are thinking of going into mathematical physics.

The first thing I should say is that getting an interview *offer* is but one stage of the complex process of getting a job. I deal with job applications in Chapter 17, *Finding a Job*. I'll assume you have read that chapter.

Suppose that one fine day, you receive a letter or an email inviting you to come to a certain place on a certain day at a certain time for that most exciting, or perhaps worrying, experience known as a *job interview*. Congratulations. That's an important achievement and not something to be treated lightly. It's possible that a good interview could be *the* turning point in your life that's followed by many decades of employment and personal satisfaction. I know that's possible because it's happened to me.

Before I go on about interviews, there is just one thing you should sort out before any job interview. *Do you really want* ***that*** *particular job*?

That may sound crazy, but actually, it's one of the serious aspects about finding a job. Suppose you received *two* job offers on *Friday*. Call them $A$ and $B$. Let's say interview $A$ is on the following *Monday* for a job in an office in your home town earning five times as much as job B, with the possibility of secure employment with a multinational company. On the other hand, the interview for $B$ is also on the Monday for a three-year post graduate degree in mathematical physics in a University five hundred miles away, with no guarantees at the end of it. What do you do?

That's where the only person who can help you is ... *yourself*. It's really no good asking your parents or friends. They will not be the ones stuck in an office for years if you get job $A$ and come to hate it.

In a situation like that, that's where personal motivation comes into play. If you have thought out what you (and no one else) really want to do, then go for it and forget about the other job offer. Job $A$ offers one sort of satisfaction in life and job $B$ offers another sort. I cannot say which is better, because there really is no yardstick with which I can measure "better" here. All I can say is, I personally would go for $B$.

## 18.1 Do your homework

Unless you've been headhunted for a job (meaning that *they* want you just as you are), you would be well-advised to do some research on your potential next job. I usually did that for the academic jobs I applied for. It's important, because you hear of some horror stories. For instance, I once heard of a mathematical physicist who had a whole string of research students that he was supervising concurrently. I heard that he was so busy with his research, he had the following approach to supervising PhD students. On a pre-appointed day, all his students would line up outside his office and he would see them one by one with quick-fire meetings. Maybe that worked for him, but I would have hated to have been one of the ten or more students he had. Doctoral supervision should be regarded as a form of mentoring, a personal one-to-one process done with care and due diligence, not something to be over and done with on a timetable.

Here's one job interview that no amount of preparation would have warned me about the question thrown to me at the end of that interview.

## The Last Question

You'll guess from my surname that my parents came from Eastern Europe. I was born in the United Kingdom and was brought up in a partial bilingual way. I was fully English in my language and education from an early age. I've always *thought* in English, if you know what that means. For instance, when I read a sign in English, I don't need to think about its meaning: I just *know* what it means.

I can't say that about my parent's language. In my childhood, English dominated conversation between me and my parents. After all, I had constant exposure to English at school, and that swamped any effort I made to learn their language. Also, I really didn't have much interest in learning what I thought of as a difficult (for me) foreign language that would not be much use to me. In our family we spoke a functional amalgam of two languages, which was reasonable at the time.

I eventually changed my mind. There came a day when I decided to spend a year in my parent's old country, on a postdoctoral fellowship sponsored by the *British Council*.

What can I say about the British Council apart from the fact that it's one of the great British institutions that, like the BBC, the Royal Society, and Parliament, has to be preserved at all costs. I filled in my application forms, sent them away, and eventually received an invitation to go to London for an interview.

I thought I was reasonably prepared for that interview. I went to London and called at the Cultural Mission of the country concerned. I waited in an ante-room until a secretary called me. I followed her into one of the strangest rooms I've ever been in. It was large and empty except for two items of furniture. One of them was an isolated chair in the middle of the room, and I was directed to sit on it.

It was as if this was going to be a quiz show and I was the contestant going for the million-dollar question. Sitting down, I turned and saw the other piece of furniture in the room. It was an enormous, long table, behind which sat six interviewers facing me, with six more people, the secretaries, behind them. The three interviewers to my left were from the Cultural Mission whilst the three interviewers to my right were from the British Council. Each of the six interviewers held large open folders, probably full of all sorts of details about myself that even I didn't know about. The six secretaries behind them were also armed with open folders. Apart from myself on my isolated chair and the big table with the the twelve people behind it, the room was devoid of anything else.

You have to understand the political situation at that time. It was in the middle of what's called the *Cold War*. That's not a fanciful description. It was as near to a real war as it could get, short of killing millions of people. Running through the middle of Europe, north to south, was a military and economic frontier known as the *Iron Curtain*, dividing Europe into two hostile and fully armed military camps. On the western side of that frontier were the countries defended by *NATO*, whilst on the eastern side were the countries allied to the Soviet Union and defended by the *Warsaw Pact*.

Both sides were armed to the teeth with conventional and nuclear weapons. Looking back now on those times, you might get the impression that it was all just posture and politics. Nothing could be further from the realities of those days. The world had come perilously close to nuclear war in 1962 and tensions remained high for years after that. People died just trying to cross the Iron Curtain. There I was, applying for a grant to cross that barrier and spend a year in the enemy camp.

I have to say that the Cultural Mission interviewers were genuinely friendly, to my surprise. I detected no hostility on their part and they did not ask me any unpleasant questions. I think they were pleased that I was interested in visiting their country.

The interview started as follows. First, one side would ask a question, which I would answer. Then the other side would ask a question, which I would answer. Then back to the other side, and so on.

The questions started by reviewing my academic experiences and going over my research plans in the country concerned. I thought I was doing well and had committed no *faux pas*. Finally, the last questions came.

The leader of the Cultural Mission interviewers asked me: "*Why do you want to come to our country?*"

I was prepared. "*As you will appreciate, I'm not very proficient at my parent's language, so I would like to spend time improving my knowledge of it.*"

That seemed to please them. So it was over now to the British Council interviewers for their last question. Bearing in mind the fact that several thousand nuclear weapons were poised to land on each other's territories, here's what the final British Council question was:

"*If you got this grant, would you be prepared to work in an atomic energy research establishment on your return to the United Kingdom?*"

...

This was an astonishingly insensitive question, given the military tensions across the Iron Curtain right at that time. Instinctively, I felt the question was some sort of trap or test. I hesitated for some moments, realizing intuitively that this question had to be answered very carefully.

The solution came. I simply spoke my mind:

"*I would be happy to consider any job offers on my return to the UK.*"

That must have been the right answer under those circumstances. If I had answered either *yes* or *no*, one side or the other might have judged me as a risk factor in some way.

As a postscript to the above anecdote, I can add that I got the fellowship, spent a great year abroad, and was never recruited as a spy by either side at any time.

The moral here is not to be too clever in an interview. Honesty is the best approach.

## 18.2 The other side of the fence

When you think about it, interviews should not be thought of as one-sided affairs, where the person being interviewed (the *interviewee*) is under investigation by the interviewers. It can work the other way. An interview gives an interviewee a chance to see something about the organization that is doing the interviewing. It's quite possible that during an interview, you the interviewee will discover what that company's work ethic is like and what disadvantages having that job could entail.

## Tax Man

Even with perfect planning, events don't always run on schedule. There came the day when I found myself living back home with my parents, with a degree entitling me to be called *Doctor*, and unemployed. I was, in essence, "resting", as they say in the acting profession. It was not clear right there and then what my academic job applications would lead to, so under mild economic pressure (never from my parents), I applied for a job as a *Tax Man*.

I suppose the Tax people were curious as to why someone with a doctorate in high energy physics would apply to them for a job. I received an offer of a first stage interview. If I passed that, then I would go on to the second stage.

I was shown into the interview room. There sat two men, whom I can only describe as *Old Guy* and *Young Guy*, as if I was being interrogated by *Good Cop* and *Bad Cop* in some crime drama. Introductions were made, and then the questions came, in the form of scenarios. *Old Guy* posed the following scenario for me to analyse. A well-known company was saving enormous amounts of tax using a somewhat morally ambivalent technicality in the law. What should I do?

I suspected a trap. If I said that I would take repressive action against the firm, that might be regarded as unprofessional. So I said that the law was the law and if the firm was technically not breaching it, then their tax savings had to be upheld.

Then it was *Young Guy*'s turn. He posed the following scenario. An elderly, impoverished widow had failed to give full details of her meagre income from renting out some rooms to students. What should I do?

Again, I suspected a trap. If I said I would overlook her misdemeanor, then I would be in breach of Tax regulations. So I said that I would apply the standard rules to her, regardless of her age and financial situation.

That seemed to settle matters. Both interviewers seemed pleased with my answers, particularly with my reply about the old lady. I thought I was doing well.

And that was when I put my foot in it. The Tax Men relaxed and *Old Guy* asked me if I had any questions for them. I thought for a few moments, and then asked my stupid question.

"*So where do you go to socialize?*" I asked, in all innocence.

They looked at each other, puzzled. *Old Guy* asked: "*What do you mean?*"

Without hesitating, I said: "*Well, you know ... Tax Men are highly unpopular. So you wouldn't go into any ordinary pub where you were known, would you?*"

I suppose that's why I didn't get invited onto the Second Stage.

## 18.3 Think on your feet

Interviews are strange processes. It's usually impossible to predict how they will go, if you're faced with total strangers.

Some interviews are comfortable experiences. The people interviewing you can be friendly and helpful. Here's what happened when I applied for a job in a field I was totally unfamiliar with.

**Thanks Professor Jim**

There I was, an unemployed mathematical physicist, with a doctorate in a theoretical discipline, going into an interview for a postdoc in experimental physics. I had used my *machine gun* rather liberally, and this job interview came up as a result.

Of course, I tried to prepare for that interview, but you can't learn all about a subject in a week. Nevertheless, in I went, overconfident as usual.

The table was large and round. I was given a chair on one side and was surrounded by professors, administrators, and secretaries. It was rather more close contact and much less confrontational than some interviews I've had.

The interview started. First they went over my personal history, and asked me to say a few things about myself. All that was reasonable and I thought it went well. Then came the questions. They started to get technical. Finally, the killer question came. One of the professors asked casually: "*What's the frequency of the radiation in typical nuclear magnetic resonance experiments?*" I should say that the job was all about nuclear magnetic resonance, otherwise known in the trade as NMR.

That got me. I had not memorized that one basic fact up. I hesitated. I started to hum. I could feel the floor slowly start to spin under my feet, just like in that Chemistry exam all those years before. I was lost.

And that was when one of the interviewers, an elderly distinguished man, Prof. Jim, chipped in, with the remark: "*It's radio frequency, isn't it?*"

That kind man had guessed I didn't know the answer, so he intervened at the crucial moment and rescued me. Of course, I said "*Yes!*", because he had cleverly changed the question into an answer.

I like to think that it was because of Prof. Jim that I got that job. I spent two good years of my life applying my theoretical experience to good use, fitting the data I collected in their laboratory to the theoretical models I developed there.

By the way, Prof. Jim worked in a completely separate group and I never really had the opportunity to thank him.

## 18.4 Phoney interviews

Believe it or not, not all job adverts are what they seem. Suppose the leader of a research group had secured, through great effort, the funding to employ an additional researcher in their group at the post-doctoral level. In many cases, University policy will require that vacancy to be properly advertised inside *and* outside the University, perhaps even internationally. The intention there is that the best qualified candidate should get the job.

I suspect that the best candidate is not always appointed. There may be hidden prejudice against race, culture, or gender. A more likely scenario is that the leader of that group might already have someone in mind, perhaps even some bright research student in their research group about to be awarded their doctorate. It's not beyond the bounds of possibility that the interviews held with candidates from far and wide would be essentially pointless, because the leader had already decided that their favourite candidate should get the job, come what may.

What do you do if you are one of the external candidates and you find out that the interviews are pointless?

I think this does not happen often enough to worry about. Certainly, in the department where I worked, I came across only one occasion where some colleagues and I thought it might happen, and we were prepared to object vigorously if there was local favouritism. In that instance, fortunately, the local candidate really was the best qualified candidate (take it from me) and he got the job.

If you are concerned about race, culture, or gender bias in any department that's advertising a position, my best advice is to go online and look at that department's staff entries. If it's a department of sixty male staff, post-docs, and research students, and you were a female applicant (or the other way round) then you might have some reason to be concerned. Fortunately, one of the characteristics of Academia these days is that you will generally see an international, diverse mix of personnel with no clear bias. I would expect all mathematical physics departments to be like that.

## 18.5 Honesty and Integrity

Phoney interviews are given by potential employers that play by their own rules, but there are candidates who behave just as badly. The following anecdote happened fairly recently and illustrates the sort of behaviour that is deplorable.

### I'll Take the Job

One day, a brilliant personal tutee of mine, *Bright Girl*, asked me to act as a reference for her. She was applying to several research groups, home and abroad, to do a doctorate in mathematical physics.

I was delighted to reassure her that I would give her my very best references, as she was a superb student with a promising future in mathematical physics.

You may think that applying for several positions at once is strange, but in practice, it's the only way to get a job. Moreover, it's not morally wrong or dubious. Simply *applying* for a job is not depriving anyone else of that job.

It was not long before I received an email from a *Prestigious Institute* abroad. They had received her application, were greatly impressed, and asked me to give her a reference. That's standard procedure.

I did that as soon as I could and in due course, she came into my office to inform me that she had been offered the studentship in that *Prestigious Institute* and had firmly accepted. I congratulated her immediately.

About two weeks later, I received a really angry email from the *Prestigious Institute*, virtually accusing me of disreputable behaviour. The girl had accepted their offer firmly, so they then immediately binned all the applications from the many other candidates, expecting to see her in a few months' time to start her doctorate with them. But, *after* she had accepted their offer firmly, what did she do? She went to London for an interview for a doctorate in a department there, had been offered a studentship, and had accepted it on the spot. In effect, she had accepted two positions at the same time. She then emailed the *Prestigious Institute* abroad to tell them she was no longer interested, thank you.

I was appalled. At no time had she told me she had such a strategy in mind. If I had any notion, I would certainly have told her not to accept two jobs at the same time. It took all my diplomacy in my reply to the outraged *Prestigious Institute* to convince them that such behaviour was not something I approved of or was aware of.

You may have your views about this anecdote. There's the view that employers don't hesitate to treat job applicants like dirt, frequently not bothering to tell them that they have not got the job, so why shouldn't job seekers play the same game?

I understand that view, but in the context of Academia, there are consequences for other candidates when such things happen. There would have been many other fine young candidates whose applications were binned by *Bright Girl*'s actions. What *Bright Girl* should have done is to have delayed her decision about *Prestigious Institute*'s offer until the London offer, accepted one of them, and stuck to that decision. I lowered my opinion of *Bright Girl* after this incident.

My recommendation is to be honest in all your dealings.

# Chapter 19

# Sitting at the Back

*I've seen things you people wouldn't believe. Attack ships on fire off the shoulder of Orion. I watched c-beams glitter in the dark near the Tannhauser Gate. All those moments will be lost in time, like tears in rain.*

Roy Batty, *Blade Runner*

In this chapter, I discuss some incidents that involve sitting at the back of a class or lecture room. Each of them has some thing to say about the way lectures and seminars play a role in mathematical physics.

If you've read some of my anecdotes in other chapters, you'll know that I've always liked to sit at the back of any lecture theatre. It's not that I'm antisocial, but early on in my academic career I realized that being at the back gives you the best perspective on whatever is going on, particularly if there's some interaction between speaker and audience. My first anecdote says more about the lecturer than about the students, I think.

## Sleeping at the Back

On this particular morning, having no lecture at eleven, I went as usual to the Maths Department coffee room, which was a disused laboratory room we shared with Physics. I got my coffee and sat down with colleagues at the Maths table. Whilst the room was large and had several tables, Maths people (which included staff, postdoctoral fellows, and research students) invariably clustered around one particular table, whilst Physics people sat everywhere else.

It's not that the two sides were antagonistic. Far from it. We generally had cordial relations and shared the administration of some degree courses that involved both sides, such as Joint-Honours Mathematics and Physics, and of course, Mathematical Physics. It's just that research-wise, we interacted very little, since we did quite abstract stuff whilst they did actual experiments.

Just after eleven, colleague Terry arrived. He got his coffee and sat down with us without a word. It was clear he was worried about something.

Normally, Terry was talkative, but today he was subdued, so another colleague asked him what was up. This is his answer.

"*I've just given my ten o'clock mechanics lecture to the Second Years.*"

"*So?*" we asked.

"*Usually the back three rows are full of students sleeping, which I've come to expect.*"

"*So?*"

"*Just now, they were sleeping in the front three rows as well.*"

I give this anecdote to point out that sitting at the back of a class, lecture, or conference talk does have its advantages. One of these is that, whilst you can see everything that's going on, most people can't see what you're up to. That may include sleeping, or even having lunch.

I'm sure I'll attract some criticism from philosophers with my next anecdote, because it's aimed at them. There's several reasons for me doing this. I won't go into them here except for the main one, which I'll go on about at length now. It's a particularly important issue for budding mathematical physicists to be alert to.

## 19.1 Philosophy is not science

For some inexplicable reasons, a lot of philosophers think that their subject has something meaningful to say about the physical universe, despite the fact that, *by definition*, **philosophers formulate no mathematical theories and do no experiments to test those theories**. I've put that main point in bold, because it tells you why *Nullius in Verba* applies to philosophers. It's rather like me expecting my three-year old granddaughter to repair my car.

A particularly irritating fact I've observed about modern philosophers is that an inordinate number of them seem to think they have an indispensable insight into quantum mechanics and time. I have even heard one elderly, eminent philosopher, in a conference on time, telling the audience that "*Time is not a subject that should be studied by scientists*".

I think the facts and common sense support precisely the opposite view. *Time is not a subject that should be commented on by philosophers.* Time is an aspect of the universe that affects everything. Anyone who thinks it should not be investigated by mathematical physicists and experimentalists should think again. Suppose you had a complicated problem to solve, such as finding a murderer, or negotiating a peace treaty. You wouldn't allow idle speculation and gossip to influence your thinking, would you? No. You would focus on hard, empirical facts alone. You may not know this, but there are many mathematical physicists working on theories of time in one form or another (such as curved spacetime, time reversal, and discrete time), and many experimentalists investigating the potential physics of those theories. A particularly spectacular example of this process at work is *time dilation*. That was theorized by mathematical physicists such as Larmor, Lorentz, and Einstein, and has been confirmed empirically in countless experiments.

I'm going on here about philosophers (and that includes metaphysicists) because their disciplines can mislead young people. I've seen it often enough: eager bright young students signing up for mathematics-and-philosophy degrees, when signing up to a mathematical physics degree would serve their intellectual needs far better. My best advice is that young people should be careful about taking what they read and hear *from anybody*, including me, as fact or "truth". Perhaps you've read about great philosophers such as Socrates and Plato and think philosophy has all the answers. It does not. You may even have started to believe that pure thought alone can solve scientific problems. It can't. Pure philosophy and pure mathematics are *not* on a par with empirical science and should not pretend to be. Do them by all means, if they appeal to you, but know their limitations.

If you are going to be a mathematical physicist, or any sort of scientist for that matter, I would advise you to take on board what Heisenberg, one of the great pioneers of quantum mechanics, wrote about using words:

...one should particularly remember that the human language permits the construction of sentences which do not involve any consequences and which therefore have no content at all - in spite of the fact that these sentences produce some kind of picture in our imagination; e.g., the statement that besides our world there exists another world, with which any connection is impossible in principle, does not lead to any experimental consequences, but does produce a kind of picture in the mind. Obviously such a statement can neither be proved nor disproved. One should be especially careful in using the words "reality", "actually", etc., since these words very often lead to statements of the type just mentioned.

[Heisenberg (1930)]

I don't discount philosophy in its right place. What Socrates, Confucius, and David Hume gave us are useful insights into how to live as humans. None of them pretended to have insights into quantum mechanics or time.

Here's the incident that led me to the conclusion that I don't have much time for modern-day philosophers.

### It Exists

Some years ago, I attended an international conference on *Time*.

Well, that was a futile exercise, I must say. The plain fact is, some subjects attract equal measures of philosophers, cranks, and hard-headed scientists. Time is one of them. Non-scientists can talk all day about it and you can't prove they're talking nonsense. The fact that they can't prove they're talking sense doesn't stop them talking.

It was that particular conference that made me decide not to go to any more such meetings. It's all very well having an interdisciplinary approach to science, but not if some of those disciplines are emphatically based on unscientific modes of thought that violate *Nullius in Verba* at every turn.

Let me say right away that the subject of Time is very much a proper one for mathematical physicists to study. Indeed, Albert Einstein made three significant contributions to the subject: *time dilation*, *loss of absolute simultaneity*, and *spacetime curvature*. Physicists do real experiments on *time reversal*, *irreversibility*, and suchlike, that show very strange and subtle effects are out there in the Universe, waiting to be understood both theoretically and experimentally. Time is a great subject for *scientists* to study.

Back to the story. There I was, sitting at the back of an enormous lecture theatre, looking down over the many empty rows of seats, as the next lecturer got onto the podium. It had been a relatively unprofitable meeting so far. Most of the talks were from philosophers and, to be frank, unscientifically trained amateurs. Such people invariably make too many hidden assumptions about *observation*, or the processes by which experimentalists acquire information, and that's the point of this particular anecdote.

As far as I was concerned, the big problem with all of the non-scientists was that they didn't bother to discuss in any way how things are measured or observed. They just assumed it's always possible. That assumption is the hallmark of "classical" thinking, which just doesn't stand up when it comes to quantum mechanics.

So on that particular morning, having had a surfeit over the preceding days of *blah-blah physics* from philosophers, I was in no particular good mood to listen to any more clap trap. And that's just when this philosopher jumped onto the podium for his talk. He started to go on about particles and suchlike, which immediately triggered a concern in my mind. You see, I was trained in quantum field theory, and there the notion of a particle is very subtle. Here's a quote from a paper on the subject:

" *Theoretical developments related to the gravitational interaction have questioned the notion of particle in quantum field theory (QFT). For instance, uniquely-defined particle states do not exist in general, in QFT on a curved spacetime. More in general, particle states are difficult to define in a background-independent quantum theory of gravity.*"
[Colosi and Rovelli (2009)].

The philosopher rambled on. I got more and more irritated with his mix of philosophy and physics words... *existence ... particle ... supervene ... space ... eternalism ... time... blah, blah, blah.* Then came the break point. He said: "Suppose a particle exists at a point..."

That was it. I had had enough. So I raised my hand to ask a question (which you're allowed to do normally) and shouted out " *What do you mean* ***exists****? Do you mean that a particle state has been prepared by some observers and they are going to detect it in some way?*"

All I got from him was a puzzled look, which meant he couldn't see what I was driving at, and he repeated: "*It exists!*".

I was baffled. "*It exists!*" is an example of *weasel words*, meaning they tell you nothing. It's a blah-blah concept. So I pressed the point. "*If you're taking about a particle at a point, you should make clear what the basis is for knowing that there's a particle there. What apparatus would be monitoring your particle? ...*

All I got from him was the repeated statement... "*It exists! It exists!*"

And that was the precise moment when I lost faith in all philosophers except three (Confucius, Socrates, and David Hume, who had a sensible view of knowledge), and promised myself never to take them seriously again.

There's one last anecdote (about sitting at the back) that I have one or two qualms about giving here. On the one hand, what I saw was incredibly funny. On the other hand, the characters involved are still alive (I hope), so may not quite see it that way. But I was the only person in the Universe who was in any position to see what happened, and if I don't mention it here, "*All those moments will be lost in time, like tears in rain.*" [Hauer (2008)].

## That Sinking Feeling

When I started my doctorate in high energy particle physics, it was the beginning of an exhilarating period in my life. One particularly important part of that life was coming across really smart, brilliant mathematical physicists.

You should know that, just as in any football team or orchestra, a sub-department of mathematical physicists will have a broad spectrum of personalities, with an equally broad spectrum of talent and reputation. Each of those mathematical physicists will be excellent at this or that branch of the discipline, but it generally falls to just a small percentage, as in any field, to be *superstars*. In mathematical physics, such a person will be recognized for creating, usually single-handedly, some great theoretical idea that has boosted the subject enormously in some way. Einstein, Dirac, and Schrödinger were such superstars.

At the time of my doctorate, I think it fair to say that there was only one mini-superstar in our sub-department. The legendary Stephen Hawking worked in the same building at that time but was in a separate sub-department associated with gravity and cosmology. Now *he* was a real superstar, his name being associated forever with the thermodynamics of black holes. The mini-superstar I refer to had written a fundamentally important paper at an astounding early age. The ideas in that paper had been expanded by others and eventually incorporated into the relatively successful theory of modern particle physics known as the *Standard Model*. I'll refer to this mini-superstar as *Our Resident Genius*.

Life in the department had its routine. On Tuesday afternoons there were local seminars, whilst on Thursday afternoons there were non-local seminars. Local seminars were talks given by staff and students from our sub-department, whereas non-local seminars were given by visiting scientists. Usually, such a visiting scientist would arrive in the morning, give their talk, and then go back home or to the airport in the evening. I recall that there were quite a few non-local talks that were given by well-known, near legendary mathematical physicists. As with all groups, some of those speakers were excellent and you learnt a lot. But there were some lemons, on occasion.

On this particular occasion, I can't say the speaker was a lemon. He really knew his stuff. It's just his *delivery* that was the problem.

I need to set the scene if you're to properly understand what happened at that talk.

First of all, the seminar room was bad. I mean, really bad. It was flat, meaning not tiered as in a proper lecture theatre. There were ten or more rows of seats, rather like benches, with an aisle running down the middle to the front. Those benches had long horizontal desk-like tops for people to write on.

The second problem was that there were many more seats than people in our sub-department, so there were invariably empty spaces, even when there was a popular talk. On this occasion, it was not a popular talk.

I turned up just before it all started. By that time, most people were already there, in their traditional distribution. Right on the front row were the Professor, Readers, and lecturers. Immediately behind them were the Post-docs (the people who did advanced research in the department on one, or two-year contracts after getting their doctorates). Behind them were two or three rows of research students (people like me doing their doctorates). Then behind them was the *void*, the empty rows of unoccupied seats, all the way to the back.

I stood by the entrance, which was right at the back, and noticed this distribution. I wanted to sit at the back and that was going to be fine. Except for one thing. *Our Resident Genius* was sitting on the extreme *right* of the back row, as usual. It was something I had noticed before. Of all staff members in that sub-department, only *Our Resident Genius* tended to sit on the back row. For some unaccountable reason and quite independently, the two of us seemed to prefer *sitting at the back*.

He was busy scribbling something, as usual, so did not notice me. I looked to my left and saw that the seat on the extreme *left* of the back row was empty, so I sat there. There were only the two of us on the back row. From my position in the left-hand corner, I could see absolutely everything. When I glanced to the opposite back corner of the room, on the right, I could see *Our Resident Genius* working away at his next piece of mathematical physics. I've noticed on many an occasion in seminars that some people keep working away on their calculations regardless of what's going on at the front. I think that's fair play. Time is precious and it's up to a speaker to create interest in their talk.

I'll call the visiting speaker on that particular day *Visitor*.

Now *Visitor* was an energetic speaker, and part of that energy involved interacting with his audience. Unfortunately, his interaction was of a rather direct form. He *asked direct questions to specific people in the audience.*

He started his talk and began to cover his ground. Then he came to a specific point, asking the question "*What is zer centre of zee group $SU(32)$?*". Now I don't need to explain any of that. I can reassure you that this was the specific point he was talking about, because it is burned into my memory. (By the way, the answer is given by the rule that *the centre of the group $SU(n)$ is isomorphic to the cyclic group $Z_n$*, so the answer *Visitor* was looking for was $Z_{32}$.)

The problem was, *Visitor* didn't just ask his question to thin air. Oh no. He turned from the blackboard and, pointing his finger directly at them, addressed the first person on the front row, to his right: "*What is zer centre of zee group $SU(32)$?*"

The staff member cringed but gave no reply. Silence. Seconds passed and you could almost feel the expansion of the Universe. *Visitor* shrugged, turned to the next member of staff and asked the same question. No reply came again. *Visitor* shrugged again and turned to victim number three. As he slowly worked his way across the front row, it became abundantly clear that he was going to interrogate every last member of our sub-department until he had got an answer from one of them. At one point, he blurted out, loudly, "*What? You don't know? Any second year research student in my department could answer zis!*"

It was like watching the Titanic sinking. A great catastrophe was unfolding before my eyes. I could see members of our staff, the ones that he had not yet reached and interrogated, cringing. Behind them, the post-docs were getting agitated. *Visitor* was going to show them all up in front of their own students!

And that's when I understood what made *Our Resident Genius* the genius he was. In amidst of the intellectual carnage going on at the front, I glanced across to the right hand back corner of the room, where sat *Our Resident Genius*.

It was then that I saw something that no one else in the entire Universe could have witnessed. I saw *Our Resident Genius* lower himself, silently and smoothly, flat horizontal onto that back bench. He just disappeared out of sight of *Visitor*, who would surely have asked *Our Resident Genius* that infernal question. If he had seen him, that is.

A few points about this last anecdote.

**1.** I think *Our Resident Genius* is a genius because geniuses solve impossible problems. *Our Resident Genius* solved his problem brilliantly.

**2.** I imagine *Our Resident Genius* knew the answer to *Visitor*'s question. But who wants to be treated like a student in public?

**3.** When the seminar was over, *Visitor* went off to the railway station to go home. The audience trooped off to the coffee room and had what could only be described as a funeral wake. I sat with them quietly and listened to the intense cursing from the staff at how *Visitor* had conducted his seminar and shown them up in front of their students. *Our Resident Genius* was not there, I think.

**4.** Many years later, I was at a conference in Dubna, Russia. I was once again at the back of a vast, banked lecture theatre with a packed international audience in front of me. Lo and behold, one of the star turns was good old *Visitor*. He had not changed. Once again, he started asking direct questions to specific people in the audience, and getting much the same response as all those years before. Some people never change.

**5.** Such experiences go to show that mathematical physics is a great subject carried out by really clever individuals, who are essentially very human, with all the positives and negatives everyone else has.

# Chapter 20

# The Administrators

Demosthenes: *You possess all the attributes of a demagogue; a screeching, horrible voice, a perverse, crossgrained nature and the language of the market-place. In you all is united which is needful for governing.*

[Aristophenes (424 B.C.E.)]

This chapter is all about *administrators*. You might have thought that that topic would have nothing to do with being a mathematical physicist. Don't be too sure.

Here's an example where academics cannot compete with adminstrative intransigence. It really happened.

### The New Building

When I joined the Maths department, it had long shared a big building with Physics. The two sides had friendly relations and we collaborated successfully over the years in running our Mathematical Physics degree.

But, as with all things, the building eventually became less than fit for purpose for us all. Physics needed more space for their experiments and Maths was expanding. So space was an issue for both sides, and there was a need for a different layout as far as Maths was concerned. Maths students should have space to walk around, meet friends, sit down and work out their problems drinking coffee over tables in pleasant areas, with whiteboards everywhere for scribbling calculations on. It doesn't help if students have to move from one grubby office to another down dimly lit corridors.

Eventually, one morning, in the coffee room, I found myself sitting next to our head of department, *HD*, who was a pure mathematician. *HD* never created stress and was quite approachable. I count my time there when he was head of department (it changed every three years) as one of the best times for me personally.

As I drank my coffee, I listened to my colleagues discussing *HD*'s scheduled visit to the main administrative building later that morning to discuss the building problem. The department had finally geared itself up for a request to the University for a brand new building, just for Maths.

It had not been easy. A whole load of reasons had to be given and justified, such as the anticipated increase in student numbers, the extra staff that were coming in due to more income being generated from public financing bodies such as the various Research Councils, as well as Commerce and Industry, and so on. *HD* was confident. Being a pure mathematician, he had all his facts ready, all logically laid out. Surely Administration would agree.

Coffee finished, we all got up, and carried on with routine departmental life. *HD* left to meet Administration and the rest of us got on with departmental duties. Lectures came and went. Lunch time came and went. More lectures came and went. Then it was quarter to four in the afternoon and coffee time once again.

We all sat there in eager anticipation of great news as *HD* joined us with his coffee cup in hand. He had had his meeting with Administration. He sat down, quietly.

A colleague asked: "*How did it go? Did you lay out the case for us to get a new building?*"

"*Yes*", said *HD* quietly. "*I laid out everything for the Finance Committee. All the details, the justification, the irrefutable logic.*"

"*And?*"

"*They just said* ***no.***"

That was the first time I experienced the reality of administrative intransigence. Here's another true story, where the sheer deviousness of administrators is revealed.

## Funding

Now you may think that being a mathematical physicist is a cushy sort of life. Well, generally *yes it is.* But there are some irritations to the job.

Perhaps the main one is finding the time to just *think* out your flashes of inspiration, to flesh out the details. But there's obviously a price to pay somewhere for that leisure and privilege.

I recall that when I started my doctorate (PhD), I was on my way to the Porter's Lodge of my College to pick up my keys, when I read a memorable saying written in chalk on a college wall by some philosophy student (or so I imagined at the time). It was the advice; "*Don't go around with your mouth open, hoping for a roast chicken to fly in.*" That's a pretty good piece of advice for any budding academic. Don't expect the academic system to give you the freedom to do your research just like that. You're going to work hard to get it, somehow.

One of the main ways of getting the resources (money and time) to do your mathematical physics is to apply to funding agencies such as national Science Councils for a research grant.

I've always been amazed that such organizations exist. I mean, how strange is it that a group of people would spend all their time on committees, deciding how to *give money away.*

But they don't just give it away. Oh no. If you want some, you've got to work at it. First, you come up with a research plan that might be of interest to other people. Your plan has to be fully costed and a detailed timetable set out. You've got to justify all your necessary expenses, such as conference trips, publication costs, University (ripoff) charges, extra staff hire, and many other details as well.

All of that can take an inordinate amount of time. Some academics are really good at it, getting enormous grants of millions of dollars, pounds, or Euros. As for myself, I was never really good at that side of business. I always found it far too diverting from the real side of mathematical physics, which is just trying to think about the research, rather than the funding of it. You may find, if you go into mathematical physics, that success is usually measured by either you coming up with an Einstein-level theory, or else getting a million plus grant. I have noticed that these two objectives never seem to have anything to do with each other.

One day, I was in my office in the Maths department, when the door burst open. Fortunately, the hinges held and the door did not fall off. What I saw was a totally furious senior colleague of mine, standing there in the corridor, red in the face.

He came in. He needed to offload his great anger. Not at me, but via me. As he calmed down, he held up a letter he had just received. It was from one of the Science Councils that funded mathematics research.

And why was he so angry?

The letter was a rejection. His funding application, based on the research he had been conducting and publishing in for years, had been rejected.

*So what*? you may well ask. People get rejected all the time.

The source of his incandescent anger was in the actual wording of the reason for rejection. It was in a little rectangular box towards the end of his letter.

I read it. If it had said "*This is a badly thought out proposal*", or even "*This proposal is based on flawed principles*", a person could at least argue with those replies. So what did the actual rejection say?

It was a beautiful example of the power of language. It didn't say *your project is useless.* It didn't say *your project is good but we don't have money for it right now.*

It simply said: "**Not urgent.**"

I don't know how anyone could fight a put-down like that. Moreover, it did the job. *No money for you!*

# Chapter 21

# Publish or Perish

*Publish and be damned*

attributed to Arthur Wellesley, (Duke of Wellington)

This chapter is about something all mathematical physicists need to do as a matter of routine, which is *publish* scientific papers and books. If you want to become a mathematical physicist, or any other sort of scientist or mathematician for that matter, you will have to attend to this activity.

There are several reasons for publishing, some of which are noble and high minded, whilst others are down to earth and basic. I'll discuss these reasons in no particular order.

## 21.1 Reputation

Like it or not, just about every human activity is affected by *reputation*, or what other people think of you. If you have a good reputation as say a sports person, you will be invited to play for some teams. If you have a bad reputation in sport, then the saying *Don't call us, we'll call you* applies.

In mathematical physics, as in all other scientific and mathematical disciplines, you will find that reputation is a fragile thing. All it takes to get a bad reputation is for your work to get discredited in one form or another. For example, suppose you published a paper or book using ideas and words taken literally from other people's papers and books, without acknowledging that fact. That is known as *plagiarism*. If you have to write essays at school or University, you will know all about plagiarism, as your teachers and lecturers will have sophisticated technology these days to catch you out.

Getting a good reputation will rest on you developing and publishing sound work. Your published ideas will be tested by other theorists and hopefully, no mistakes found in your calculations. If your theory helps others to create new theories and experiments, you will be acclaimed. Your name may even be associated with constants of physics, such as *Planck's constant* in quantum mechanics. You will be acknowledged as the creator of that theory, because the evidence that you did it will be in the published literature. I can go to library archives and find the original journals that published Planck's research in 1900, and confirm for myself that he was the first person to think of the quantum of energy and use it to fit real experimental data.

Mathematical physicists with great reputations have those reputations because their theories are useful even perhaps centuries after they were first published. We celebrate Isaac Newton not because he was such a great guy, but because so many of his ideas actually work and are useful even today. His personality and beliefs are of no significance in that respect. The same goes for other great mathematical physicists, such as Maxwell, Einstein, Dirac, and so on. Their published work is great, so we think of them as great.

In experimental science, one sure way of getting a very bad reputation is to get caught fiddling your data. In mathematical physics, you can't really do that, because it's usually easy for other mathematical physicists to go over your work and spot your errors.

There is one aspect of mathematical physics where you could get yourself into trouble, and I've seen it in real life. That's to do with the *principles* you assume when you create your theory. An example of what I have in mind occurred at a scientific conference years ago. One of the speakers tried desperately hard to convince us in the audience of the reasons why Special Relativity and quantum mechanics are incorrect. I don't have any problem with intelligent debate, but if a theorist such as the one I'm referring to cannot grasp some basic, common sense statements, then that theorist gets known as having some mental problem.

It's like arguing with believers in the Flat Earth and other conspiracy theorists. Once you see that their conditioning is permanent, there's no use in discussing anything with them.

## 21.2 Communicating

How does a mathematical physicist tell others of their ideas?

There are several ways. The most obvious one is to actually *talk* to people face to face. That may involve a discussion with a colleague over coffee during a break in your department, or a visit, possibly to another country, to talk to a particular individual who can discuss your ideas with you.

That's not always possible however. You could be the only mathematical physicist interested in quantum mechanics in your department, for instance. Or you might not actually know anyone outside your department to visit. But *conferences* are a way to get around that particular point.

A conference is usually (but not always) a well-organized event in some pleasant location (usually), over a day, a few days, or more than a week. At a conference you can meet lots of new people and give talks to them on your work. For beginners in any area of knowledge, a conference is an ideal opportunity for people to find out that you exist and what you do.

Conferences are great opportunities to travel to exotic places and to meet fascinating and clever people. But there are some downsides that you should be aware of:

### *Funding*

Going to a conference involves paying for travel, accommodation, and invariably, *conference registration*. All of that adds up. Fortunately, there are organizations that seem to have money to throw away, so it's usually possible to get finance for a conference trip. But getting finance does involve filling in forms to justify why you should get all that money. That was something I always hated to do.

### *Predatory conferences and journals*

Believe it or not, there are bogus (fraudulent) conferences set up to relieve inexperienced academics of their money. Do be careful in this respect. I regularly get invitations to conferences, but most of those usually turn out to be scams. Recently, I was invited to take part in a conference on a cruise ship. You can imagine that the scam centred on the payment of a large sum of money for the cruise itself, as well as paying for conference registration and suchlike.

The same things happens a lot with journal publishers these days, unfortunately. It's all too easy for a fraudster to set up a web-site listed as "*International Journal of Mathematical Bricklaying and Applications*" and promise quick publication in their online journal, for a fee of course.

Before you part with any money for a conference, do check it out as thoroughly as you can online. You can find web-sites that give lists of known bogus conferences and their organizers, if you look hard enough. Be especially careful if you receive an invitation to what looks like an important position, such as acting as chair of a meeting. Making their victims feel important is one of the particularly effective ways these tricksters have. The best way to avoid such traps is to follow Socrates' advice and *know yourself*. If you don't have an obvious international reputation yet, but are being invited in an email from someone you never heard of as if you had, there's a good chance your ego is being worked on.

In the long run, it is better *not to publish at all* rather than publish in a predatory journal. If you do get published in a bogus journal, that may haunt you for the rest of your career. I know some people who did just that and I would not now collaborate with them even if they paid me.

### *Your reputation*

Going to a conference and giving a talk on your work is not risk free. You could give people the impression that you are a bad theorist if you give a lemon talk. If you go to a conference, make sure that what you talk about is sound. A conference is after all a public forum, and *you* are on stage under close scrutiny.

Having said that, I would say that a conference is a great starting point for a beginner. I've seen young theorists give the first talk of their careers in public, and I've come away impressed with their delivery, their sparkle, their youthful enthusiasm, and the sheer brilliance of their ideas. Young people are the future, after all.

If you are unfortunate enough to meet stupid critics that give you a bad time during your talk, *do not get upset*, provided you are confident of your material. You will survive if your material satisfies *Nullius in Verba*. That's the best anyone can do. We all meet people we cannot convince, sooner or later.

### *Overdoing it*

I went to *four* conferences and visits in one particular year, all involving travel to other countries. Believe me, after all that, I felt as if I wanted to burn my passport. You can see too many airports. Moreover, you don't really get much peace and quiet when you travel. Sure, I find actual motion, such as by train, very stimulation. I tend to think of mechanics at such times, but it not easy to do calculations on trains or aeroplanes unless you are prepared, perhaps with a *mobile office* of the sort I discuss Chapter 15.

## 21.3 Legacy

Why is publishing written words so important? Well, it's to do with *legacy*, the memory of you that you leave behind when you've moved on in the *Great Scheme of Things*. There are two sides to legacy: I call them *persistence* and *priority*. I'll discuss these two points next.

### *Persistence*

Have you ever heard of the *Library of Ashurbanipal*? Go online and be amazed, inspired, moved, impressed, ... It's the oldest known surviving royal library, with over **thirty thousand** clay tablets recovered so far. My point is simple. When the electricity generating stations go off-line, your computer, tablet, and smart phone gadgets will *all* become so much junk. Ashurbanipal's library will still be there.

There's an even more insidious, long term problem. Hardware and software become out of date very quickly and many old computer files become unreadable. When I first started in the maths department, all our computer files were stored on enormous tape reels, precisely the sort you see in dated science-fiction films, running on computers the size of elephants with half the processing power of any smart phone. Those tapes are now essentially inaccessible to anyone except specialist firms charging huge amounts to read those tapes.

More recently, a whole load of "archives" that I saved on my personal computer more than ten years ago in the now-not-then obsolete fax format (computer files ending in "dot awd") are unreadable on today's systems[1], unless I pay for specialist programs to read them. I would have been better advised to save my archives on paper, or even on clay tablets. Not that long ago, many scientific articles were saved in a format known as *postscript*, but nowadays finding software to read that format has become difficult. Currently, the *pdf*[2] format is used extensively for sending scientific articles over the internet, but I dread to think what will happen when that format become unreadable.

On the other hand, articles written by hand on clay tablets for Ashurbanipal's library more than two thousand years ago are still around and quite readable (translating them is another issue entirely). *That's* persistence. That applies not only to clay tablets. I've still got one or two of the real books that I had as a child, including an exercise book of crayon drawings I did in nursery school at age four. I won't say how many decades ago that was.

## Library Archives

One of the great, not to be missed, experiences in life is to step into an old, dusty library. Seeing shelves of leather-bound tomes, old journals, and ancient books, always sends shivers down my spine. These are the persistent memories of the men and women who came before us. It inspires me think of time and its physical relationship with information. What is a library other than a method of preserving *ideas* against the ravages of time? Libraries are no more and no less than artificial *memories*. It is an absolutely crucial fact that without our memories, we humans are nothing, because we are defined by our conditioning, and conditioning depends on memory.

I remember one time I had to find an old science paper about the *Bohr-Sommerfeld atom*, a now obsolete theory of the atom developed a few years before the tremendous explosion in mathematical physics that gave rise to *quantum wave mechanics* (the *Schrödinger equation* and all that) in 1926.

I went to the main University Library desk and asked for the key to the basement. Then I went down some grimy stairs, unlocked the door to the Archive Room where all the old journals were stored, and went in.

It was eerie, dark, and silent except for my footsteps. I was quite alone. I switched on the lights and saw the gloomy rows and rows of ancient and forgotten books, journals, and dissertations, plus one or two trays of rat poison on the ground.

I consulted the index book and found out where the journal was that contained the article I wanted. I worked my way through the maze of shelves until I found it. I took it from the shelf, shook off the dust, and tried to find my article. But I could not! Many of the pages were still *uncut*.

You may not know about this, but in the old days (maybe a hundred years ago), many books, magazines, and journals would be sold with their pages uncut. Books were made by printing several pages of text onto one big sheet of paper, folding that big sheet as required, and then sewing lots of folded big sheets together to form the book. Quite frequently, book binders would not bother to cut the folds at the edges of the book, so adjacent pages would be stuck together.

Finding that physics journal (which was quite massive, having well over a thousand pages) uncut made me realize that *no one had tried to read that particular book before, over the last eighty-odd years since it was printed.*

Here's what one book-lover writes about this: "*If you find a book with uncut pages, you can be sure that it has never been read.*" [Cook (2010)]. In such cases, it was expected that the first person who wanted to read the book would cut the pages with a knife.

Finding an uncut book like this is like an archaeologist digging up a clay tablet from Ashurbanipal's library: they would be the first person to have looked at that tablet for a couple of thousand years.

### *Priority*

No one likes others to get credit for their work. That's only natural, given that society can give generous rewards for personal achievements (such as knighthoods for cycling up mountains faster than other people, or lordships for giving shed-loads of money to political parties, or losing millions at a bank).

It's no different in science. Science is done by humans, after all. It's nonsense to think otherwise. If you've spent twenty years constructing the *Theory of Everything*, you would be annoyed to say the least to see that someone else had published the same theory a couple of days before you did. What would have happened would be that *they* had established *priority* in the matter. Now and forever, history would record in its books that *they*, not *you*, were the discoverer of that theory.

In order to claim priority, most scientists would, in the days before the internet, publish *preprints*. These were invariably flimsy documents explaining in detail their results. Preprints would be sent out by regular mail to other researchers in the field, all over the world, *before* that research was sent to a publisher for refereeing and hopeful inclusion in a good journal.

The preprint system was necessary for establishing priority, because the refereeing process at a journal would often take time before the editors of the journal would accept the article. The reason is that science requires verifiable work. If experimental results or theoretical calculations are unsound or unverifiable, then they should not be published. A journal that publishes bad research loses its reputation. Checking submitted research properly requires finding good *referees* (experts in a given disciple who read submitted articles) to check submitted papers and then give their opinions about them. All of that can take a great deal of time. In mathematics, important papers may take several years to pass the refereeing process.

One of the big problems for editors of journals is finding good referees, because there is a paradox here. Let's say someone had just discovered a completely new theory and submitted it as a paper to a well-known science journal. It would have to be checked thoroughly by an expert. But as a completely new theory, there would be no experts around yet except for the author.

Most submitted papers are not, in fact, startlingly new or original. Most of them are either incremental developments of recent theories, or plainly bad papers. Refereeing those sorts of papers is a chore and it gets harder and harder to find referees for papers based on unusual or controversial research.

In my time, as a postgraduate student, I was often roped along with the other postgrads into collating pages of other people's preprints and stapling them together. Usually, there would be several *hundred* copies of perhaps three or more preprints to be made up *every week*, ready to be sent out world wide by post (most of them by airmail). All of this was done to establish priority.

In the department where I got my doctorate, there was a dedicated room for *incoming* preprints that outside researchers has sent us. On one side of the room were the display racks, where every Monday, the hundreds of preprints from other universities and research institutes that had arrived over the weekend would be set out. Below the racks were the filing cabinets containing all the preprints that had arrived over the last year or so. Of course, once a paper had been published in a regular journal, the original preprint would be regarded as out of date and usually destroyed.

There was fierce competition amongst colleagues to get at those preprints, as people in the department were eager to keep an eye on priority. Everyone needed to see how close other people in the world were to what *they* were researching on. Priority has always been a big factor in science. For example, one of the well-known controversies in science was the argument as to who discovered calculus first, Newton or Leibniz[3].

### *The Internet arXives*

These days, things are much simpler than when I started in mathematical physics. Instead of sending hundreds of real, printed paper preprints around the world by regular and air mail, mathematical physicists, like other scientists and mathematicians, post their preprints on internet collections of like-minded articles. There is an extraordinarily useful internet depository that is freely accessible to anyone, including you, known currently as the "arXiv". The spelling is deliberate: the "X" is not meant to represent the letter X but the Greek symbol $\chi$, which is often pronounced "Ki".

Typically, when you're about to submit your finished paper to some regular publisher, you now post a copy on the arXives. It will be moderated, meaning looked at briefly but not checked thoroughly, just in case it is plain loopy, off-topic (such as discussing sociology), or potentially a spoof article. More than **ten thousand** articles in science and mathematics are posted each month on the arXives. They are usually never taken down, so they remain as a permanent statement of priority, even if the actual article never gets published in a regular journal.

As I said, the arXives are open to everyone to download articles. You do not have to be an academic. You do not have to pay or register. It is a triumph of academic freedom and science, available to all. If you want to see the latest research on say cosmology or quantum optics, you can find it on the web site *https://arxiv.org* and navigate your way around the vast number of articles dating from the year 1991, in a huge range of scientific fields.

If you have become interested in what mathematical physicists do, then I'd advise downloading some articles from say the *hep-th* section (high energy particle theory) of the arXives I use. Those articles are best downloaded in pdf format, in my opinion. They will show you what properly produced scientific papers look like, with Titles, Abstracts, Author affiliation, Sections, Conclusions, and Bibliography.

The day may come sooner than you ever expected when *you* will have the unbeatable pleasure of seeing an article you have written getting posted on the arXives. Just make sure that you do it properly, as those articles are going to be in public view for a very long time. Whilst it is true that everyone can *download* any arXive articles, the system there takes some care in making sure plain rubbish is not *uploaded* to it. Unfortunately, there are people who are not properly trained in science who would use the arXive for scientifically unacceptable purposes, such as politics or religion. Therefore, in order to post an article on the arXive, you need to establish some reasonable credentials. You can find details on the web site I give above.

## 21.4 Submission

Years ago, when I first started to write scientific papers, I had no real problems. I didn't write vast numbers of papers, but usually I managed to get them published in reasonable journals.

You should understand how the traditional system of research publication works. It's designed to uphold the principles of *Nullius in Verba*.

First, a researcher (or a group of researchers) would work over a period of time on some previously under-researched problem. Some problems would be well known to the scientific community, but other problems might be devised by the researcher themselves. Quite often, the researcher would spend months or even years working out the details of their work. Before the days of computers this would involve many pages of notes and calculations, all by hand.

Eventually, the researcher would have enough material to write a "paper". That's shorthand for a written document structured in a standard pattern that's universally accepted as the most useful way of doing things. At this point, a researcher becomes an *author*.

A "paper" could be a single page or twenty or more pages long. There would be a proper title, the author's name and address, an abstract (a brief summary of the problem and the solution given in the paper), all followed by a number of sections. The first section is usually be called *Introduction*, and in it, the author explains the problem, how they solve it, and the conventions used (such as what system of units employed).

The sections following the Introduction discuss the theory and/or apparatus used, give the results of the research, and frequently, in mathematical physics, how those results were calculated. This is particularly important in the case of new theories. The idea is to allow other people to understand the calculations. There should be no mysteries in mathematical physics, no hiding of techniques or deliberate confusion designed to cover sloppy work.

The last section of a paper usually contains an overview, the conclusions of the research, often some vague promise to do more work in some related area, and then perhaps thanks and acknowledgements to various people who had helped along the way. At the very end of the paper we should find a clear list of *references*. These contain all the information needed to help readers to find for themselves the articles and books the researcher had referred to in their paper.

## 21.5 References

It's very important to give full references for two reasons.

**1.** No one likes an author who seems to have used other people's work but failed to refer to those other people. That always comes across as unfair and self-promoting. Don't do it.

In passing, I should say there is an ongoing question concerning Albert Einstein himself. He became world famous because of a particular paper, the now-famous 1905 landmark paper on Special Relativity [Einstein (1905)]. In that paper, he invoked the principle that the speed of light should be the same in all inertial frames of reference with standard coordinates. However, there is no reference in Einstein's paper to the crucial experiments of Michelson and Morley that established that fact [Michelson and Morley (1887)]. It seems, however, that he was fully aware of developments in the field [Isaacson (2007)]. Make up your own mind.

**2.** The second reason is *accountability.* Suppose I wrote in a paper that an experiment had shown that some particles were repelled by gravity rather than attracted. That would be sensational, if true. No one would believe it, however, unless I had supplied full references pointing towards that experiment. *Nullius in verba* applies very much in scientific papers. Those references should give enough details for readers of my article to investigate my claims and find out if they were true.

## 21.6 Peer review

Don't confuse *references*, discussed above, with *refereeing*, which is the *peer review* part of the process of getting published. Here "peer" means "your equals", meaning other theorists who will look at your work open-mindedly, carefully, and impartially. Peer review is in principle the best way for science and mathematics to publish good quality research keeping to the principles of *Nullius in Verba*.

Having submitted your paper to a reputable journal, assuming you have found one and not a predatory journal, you the author will have to wait for a verdict from the editor of the journal. Papers do not get published these days just like that, unless they are sent to *predatory journals*. These bogus publishers will publish any old rubbish as long as the authors pay them. You can be sure that predatory journal refereeing is either minimal or non-existent with such tricksters.

The Editor who first receives your submitted paper will usually assign two or more separate and independent *referees* to go over your paper as carefully as they can. If they report that they agree with your work, then the Editor will accept your paper to be published in a forthcoming edition of the journal. If one or either of the referees do not like your paper, then you may be asked to rewrite it, or go away and submit it to some other journal.

All of this, when it operates properly, is sound science. It's actually sometimes in *your* best interests that referees reject your paper or spot errors. You should not want any of your work to be published unless it's good work. Once in print, any errors that there are will be there for countless people to see for generations, perhaps hundreds of years. As I've indicated in previous chapters, science cannot be fooled in the long run. Sooner or later, your mistakes, sloppiness, or cheating will catch up with you. The best policy is to strive for the very best standards and work with referees rather than think of them as enemies.

## 21.7 Rejection

Sooner or later, you will get adverse comments from referees (*Everybody gets rejected, sooner or later. See below for details of a notorious incident involving Einstein*). Editors usually send referees' comments to authors, because rejection need not be the end of the story. Incidentally, referees are usually anonymous. You will not know who they are.

If your paper is really bad and cannot be improved, then you may have no choice but to give up. However, my experience has been that referees are themselves frequently wrong. If you read their comments on your paper carefully, you may spot some assumption they have made about your work that is wrong. In such a case you can get back to the Editor and point out the referee's mistake. There is a *backstop* available in some journals known as the *Adjudicator*. This is some senior professional who can be appealed to in a case of a legitimate dispute between an author and the referees.

This process works. I once had a series of four related papers on discrete time published in a reputable journal. Every one of those papers had to go to the Adjudicator, because the referees had raised incorrect points and I could prove that they had misunderstood what I and my research student Keith had actually written in our papers.

The psychological shock of rejection cannot be ignored. When it comes, it hits you hard. I think this is especially true in mathematical physics because the subject is in such a special position. As I've pointed out before, I think of mathematical physics as virtually an art form. It uses the rigour and power of mathematics to model the intuitive concepts of physics, based on limited or even non-existent evidence from the laboratory. That's asking for rejection.

Suppose you've found a wonderful mathematical model that is consistent and sound. Nevertheless, a referee may reject your physical intuition, something that would not happen in a purely mathematical discussion. I once had a rejection based on one referee's opinion, which was one line long: "*The chronon (unit of discrete time) is not a legitimate physics concept.*"

In that instance, I was foolish enough not to dispute rejection on that crass basis, and sent my paper off to another journal, where it eventually got published.

There have been famous, Nobel Prize winning mathematical physicists who have had controversial rejections. Two cases spring to mind but I'm sure there have been many more.

### *Einstein and gravitational waves*

Einstein developed his theory of General Relativity and it became accepted by theorists. However, important conceptual issues remained. Surprisingly, Einstein himself seemed confused initially about the idea of *gravitational waves*. These are disturbances in the distance structure of spacetime that travel at the speed of light.

What happened in 1936 has been reviewed thoroughly by Kennefick [Kennefick (2005)]. Einstein and a collaborator, Rosen, submitted a paper to the journal *Physical Review* entitled "*Do Gravitational Waves Exist?*". That journal was then and remains a leading science journal. Einstein and Rosen said in their paper that such waves did *not exist*.

Einstein's paper was refereed and ... rejected. Yes. **Albert Einstein's paper, asserting that gravitational waves do not exist in the very theory that he had created, was rejected by an anonymous referee.**

Einstein could not accept such a decision. It was not simply a matter of believing that he was correct and the referee wrong. Einstein was simply not used to being refereed. He wrote to the Editor of the *Physical Review*:

```
Dear Sir,

We (Mr. Rosen and I) had sent you our manuscript for
publication and had not authorized you to show it to
specialists before it is printed. I see no reason to address
the- in any case erroneous- comments of your anonymous
expert. On the basis of this incident I prefer to publish
the paper elsewhere.

Respectfully ...
```

Apparently, science journals in Germany during the time Einstein published there, before he moved to America, tended to have a different policy regarding peer review, generally adopting the attitude "*better a wrong paper than no paper at all*" [Kennefick (2005)].

In fact, Einstein and Rosen had made a conceptual error, confusing genuine singularities in the metric field with so-called *coordinate singularities*, which could be removed by choosing better coordinates. Einstein realized his error and submitted to a different journal a revised paper that now allowed for cylindrical gravitational waves. That proved fortunate, considering that gravitational waves have now been detected [LIGO Scientific Collaboration and Virgo Collaboration (2016)].

### *Schwinger and Cold Fusion*

Julian Schwinger was a Nobel Prize-winning mathematical physicist who developed powerful theoretical techniques in particle physics. Towards the end of his life he became somewhat isolated from mainstream physics because of his refusal to follow the main herd.

Famously, he used the motto "*If you can't join 'em, beat 'em*" in one of his books on quantum field theory, where he expounded his empirically motivated approach known as *source theory*. That theory imagines theorists and experimentalists in the position of someone trying to decide what sort of animal is inside a cage in a zoo, the problem being that it is night and there is no light. By poking through the bars of the cage with a long pole, that person could perhaps find out. A gentle push *here* and a nudge *there* through the bars of the cage should elicit responses from the animal inside the cage, establishing that it is in fact a tiger and not an elephant. Likewise, by theoretically "prodding" or tweaking models of physical systems under observation carefully, a theorist could determine properties of that model that might be tested by experiments.

I've used Schwinger's source theory approach to quantum field theory and it's an extraordinarily powerful technical weapon. Used in conjunction with Feynman's *path integral* approach to quantum mechanics (mentioned in Chapter 8, *Why?*), a mathematical physicist has enormous technical power at their disposal. In 1989, the standard classification system in physics known as PACS (for *Physics and Astronomy Classification Scheme*) had an entry entitled "11.10.$Mn$ Schwinger source theory" [PACS (1989)]. *That* is the measure of Schwinger's importance.

In 1989, there was a significant breach of standard scientific publishing protocol. Two chemists, Fleishmann and Pons, announced to the ordinary (non-science) press that they had discovered *Cold Fusion*, which is the production of energy from nuclear reactions, not chemical reactions between atoms, at low (ordinary) temperatures.

There were two critically significant facts about what Fleishmann and Pons had done. First, they had not submitted their work to peer review with a regular science journal but instead had "gone public" in a sensationalist manner. That was a clear violation of *Nullius in Verba.* The other fact speaks for itself and shows how important *Nullius in Verba* is: no one has reproduced their claims about Cold Fusion. A lot of money and time was spent by experimentalists and mathematical physicists alike in ultimately futile attempts to verify Fleishmann and Pons' claims. To date, Cold Fusion remains a subject that most scientists will not touch with a barge pole.

Being an individualist, Julian Schwinger decided to look into the theory of Cold Fusion. He formulated some ideas about it and wrote them up as papers. They were rejected. One of the referees wrote:

*I have not read this paper, but it must be wrong.*

[Shah (2006)]

Regardless of the facts about Cold Fusion, that referee's comment must surely rank as one of the crassest statements in the whole of science. Schwinger was so incensed by these rejections that he wrote this to the journal concerned:

```
What, pray, in my 55 years of not unsuccessful research
justified such contempt? I submit that giving anonymity to
narrow minded specialists grants them a license to kill.

I want no more of this. Please inform whoever might be
interested that I resign as a Member and Fellow of the APS
[American Physical Society].

You will, of course, return the copyright agreement that I
signed; all rights now revert to me.

Incidentally, the PACS entry (1987) 11.10 Mn can be deleted.
There will be no further occasion to use it.

Schwinger.
```

I was fortunate to meet Schwinger and his wife Clarice in 1993, a year before he died. Here's an anecdote involving the two of those lovely people that says something about science publishing.

**Advertisement**

The year 1993 was the bi-centenary of the birth of the extraordinary mathematical physicist George Green, whom I have mentioned in Chapter 5, (*What We Do*). Along with Nottingham City Council, my University organized an international meeting to mark the occasion. First were lectures in Nottingham, where Green came from, and then in London, where a plaque in Green's honour was installed in Westminister Abbey, next to Isaac Newton's grave.

As part of the celebrations, the *Royal Society* hosted a meeting in Carlton House Terrace, their headquarters in London. A highlight was a lecture from Mary Cannell, a biographer of Green's, whose books discuss Einstein's connection with Nottingham [Cannell (2001)]. The Physics Department at Nottingham University has one of only two *Einstein blackboards* in the UK, the other being in Oxford. Einstein came to Nottingham to consult with one of his English translators and he gave a public lecture at the University. Immediately it finished, the blackboard on which Einstein had written some basic equations in General Relativity was seized for posterity.

Einstein knew all about George Green and his remarkable contributions to mathematical physics. These had for years been unrecognized in Britain whilst over in Continental Europe, Green's work had long been known and appreciated. Our celebrations of Green's bicentenary were aimed at redressing that historical oversight in the UK.

After the lecture at the *Royal Society*, an usher opened the lecture theatre doors and announced to the huge audience that the buffet was ready. An excellent meal had been laid on for us by the *Royal Society*. I can picture it now. Perhaps an analogy might give you an idea of what happened next. A *watering hole* is a geological depression containing water. Imagine a watering hole in the middle of the Serengeti National Park in Africa at the height of the dry season. Now imagine a vast herd of parched wildebeest racing for that watering hole.

No sooner had the announcement been made, and the audience was off. And I mean off. It was as if all of us had been on a three month starvation diet. A huge flood of people, including me, got up and stormed out of the lecture theatre and charged into the buffet room.

It did not last long. In about ten minutes, it looked as if a plague of locusts had passed through. The room where the buffet had been was full of satisfied audience members, finishing their plates of food and their sherries, talking at about the decibel limits of human auditory endurance. And the tables were now quite bare of any food.

As I stood there, chatting to colleagues amidst the excited throng, I suddenly saw one of the most pitiful sights you can imagine. Walking slowly from the lecture theatre into the buffet room were the elderly Julian and Clarice Schwinger, quite confused about what was going on, unaware that there was going to be no food for them. As one of the organizers at the Nottingham end, I had looked after Julian and Clarice. For example, I was their chauffeur at one stage, and took Clarice to an electrical store to buy a hair dryer. But here in London, I assumed someone else would keep an eye on them.

It suddenly hit me. None of us on the George Green Memorial Committee had thought of the London end.

I desperation, I found the usher who had made the buffet announcement and informed him that our honoured guests the Schwingers had missed their food. The excellent usher immediately understood the situation and went to organise some food for them, whilst I headed off Julian and Clarice, who were wandering alone and confused in a sea of guests, all of whom seemed to find the Schwingers invisible. They were being completely ignored yet they were the guests of honour!

I cornered Julian and Clarice by the side of the room, realizing that I had to keep them occupied until the usher returned. What to say? In desperation, I did the only thing that came to mind. I started a scientific discussion with Julian. It sounds crazy, but right there and then, I asked him about his *proper time approach to quantum field theory.* That's just one of the many brilliant ideas that Julian had come up with and worked on over the years. He really was a genius of a mathematical physicist.

As we talked, with Clarice looking on, Julian mentioned his latest research, which he said employed his *source theory.* As that usher returned and led the Schwingers to their food, Julian said that he would send me his latest work on that subject. I nodded, not expecting him to remember.

Three week later, after everyone had gone back home from the meeting, I went to my mailbox in the Maths Department and found a package. It was from Julian Schwinger. He had indeed remembered our conversation and true to his word, he had sent me his latest reprints on *sonoluminescence*.

I have three remarks to make about this.

**1.** *Sonoluminescence* is a remarkable phenomenon. There are situations where a small gas bubble in a liquid collapses, with the emission of light. Towards the end of his life, Julian was using his expertise in mathematical physics to try to understand that phenomenon. He was uniquely qualified, because he got his Nobel Prize for his contributions to *quantum electrodynamics*, which is really the quantum theory of light and its interactions with matter.

**2.** Reprints are better than the preprints I mentioned before. Preprints are versions of a paper *before* publication. Reprints are printed copies of what finally appeared in an actual journal. The reprints Julian sent me were at the height of sophistication and elegance, beautifully written, revealing a master craftsman at his creative best.

**3.** You will have read above about Julian Schwinger's falling out with a journal over the way they had refereed his papers. As Shah states in his historical review, Schwinger had to find other journals in which to publish, and mentions the journal *Proceedings of the National Academy of the Sciences* [Shah (2006)]. Schwinger used his membership of that Academy to publish papers that would not have to go through the refereeing process.

There is a twist to this. Regular journals usually did not charge authors to have their papers published. Publishers of journals usually got their costs reimbursed from science libraries that would pay large sums to get hard copies of those journals. However, some struggling journals required authors to pay *page charges*, which depended on how many pages were published. I know that *Proceedings of the National Academy of the Sciences* must have had page charges, and that Schwinger paid them.

I know this, because at the bottom of every reprint Julian sent me was the paragraph:

"*The publication costs of this article were defrayed in part by page charge payment. This article must therefore be hereby marked "advertisement" in accordance with 18 U.S.C* §1734 *solely to indicate this fact.*"

Schwinger was not just a technically superb mathematical physicist. He was a man with deep and sound scientific principles, a firm believer in *Nullius in Verba*, an individualist, and a free-thinker. Because of that, he became marginalized by a subset of the scientific community that had been conditioned into a certain rigid belief structure that, in my judgement, was inconsistent with *Nullius in Verba*. His work was aimed at explaining the world that we *actually* observe, not of what we *imagine* we observe. The difference between *actually* and *imagine* is at the heart of *Nullius in Verba*. It seems to have been forgotten by some mathematical physicists currently working in *String Theory* and *Many Worlds*.

As far as writing a theoretical paper on Cold Fusion is concerned, Schwinger was right to do so. Mathematical physicists have an agenda to investigate any new idea about the physical universe. Their papers should not be dismissed out of hand just because someone *believes* them to be wrong, as in the case of Schwinger's referee above. That's contrary to *Nullius in Verba*. Every paper should be read carefully by referees and any technical errors and inconsistencies with *Nullius in Verba* found and pointed out. If there are no problems, then the referee must declare that and vote for publication. Science cannot be done on the basis of a popularity poll.

## Chapter 22

# Strings and Things

*The scientists from Franklin to Morse were clear thinkers and did not produce erroneous theories. The scientists of today think deeply instead of clearly. One must be sane to think clearly, but one can think deeply and be quite insane.*
*Today's scientists have substituted mathematics for experiments, and they wander off through equation after equation, and eventually build a structure which has no relation to reality.*

Nikola Tesla [Tesla (1934)]

I think Nikola Tesla must have had a time machine. What he wrote in 1934, quoted above, perfectly encapsulates some big issues in current mathematical physics.

Perhaps you've been influenced by my anecdotes and enthusiasm for the subject to imagine that mathematical physics is free of opinion, bias, hype, and other base human attributes. You would be mistaken.

In my experience, mathematical physics is a wonderful subject conducted by brilliant but occasionally flawed human beings. Sometimes, mathematical physicists have gotten themselves trapped into working on topics that seem to have no empirical content, all because those people have a particular conditioned view of "truth" and reality. For example, you can still find mathematical physicists who believe in the *Hidden Variables* view of reality and reject what quantum mechanics would sell them.

It's important, indeed, crucial for any beginner in mathematical physics to be aware of the debates that are going on in the subject, because which side you're on can dictate what you do in your career. A case illustrating this is Einstein, who took a *realist* view of the Universe as quantum mechanics developed, because he was convinced that "reality" was something objective. That conviction stayed with him until the end of his life and strongly influenced what papers he published. In the previous chapter I discussed Julian Schwinger, who waged a lone campaign on behalf of his source theory.

Because having a "philosophy" of what we're doing is a necessary ingredient in any research, I don't apologize for the discursive nature of this chapter. It's going to be necessary for any mathematical physicist to have a view about what they're doing. You might as well start thinking about this as soon as you can. My task here is not to convince you one way or the other about *Many Worlds*, *Hidden Variables*, or *String Theory*, but to alert you to the need to have an informed view about such research areas.

## 22.1 The broadcast

Why did I write this book? I never planned on it. What tipped me over the edge all of a sudden was what I heard on my radio one day.

**Is this Science?**

One morning, several years, I tuned in to *Radio Four*, the flagship radio channel of the illustrious *British Broadcasting Corporation*, known worldwide as the *BBC*. Normally a reliable conveyor of news, verified facts, and wisdom, *Radio Four* let itself down on that occasion, in my humble view.

Just after the eight o'clock news that morning, there was an interview with a well-known popularizer of modern science. Let's call him *Science Guy.*

*Science Guy* is a scientist and quite clever in his own way. He looks good, sounds good, and presents his programs well on radio and on television. Usually, he is a reasonable representative of the science community and communicates science news effectively to the general public, which is a desirable end in itself. On this occasion, however, that seemed to me not to be the case. What he said almost made me choke and more than a bit annoyed.

He said that "*The majority of physicists believe in the* Many Worlds *theory*", or words to that effect.

That assertion struck a nerve somewhere in my brain. Over the next few hours, days, and weeks, *Science Guy*'s assertion wormed its way deeper and deeper into my subconscious, meeting other bits of scientific nonsense that had invaded my thoughts over the years, until the day came when the whole ugly mass of conjecture and *blah-blah physics* exploded and kick-started the processes that led to this book.

I realised that someone, somewhere, had to speak up for *Nullius in Verba*. I thought *it might as well be me*.

After listening to the popularly acclaimed *Science Guy*, I became concerned for three reasons.

### *It's a vacuous theory*

There may have been a huge number people, many of them just like you, listening in to what *Science Guy* said that day. I checked the listening figures. The *BBC*'s own statistics say that over ten million people listen in to Radio Four over a week. Given that the morning news is important and can be listened to in commuting cars, there could easily have been several hundred thousand people listening in to *Science Guy* that morning.

The vast majority of listeners would be scientifically untrained and happy to be led by the nose into believing almost anything said by *Science Guy*. None of that is their fault. Given the reputation of the *BBC*, most people would have believed that whatever *Science Guy* said was true and supported by evidence. The problem is, what he said was an unproven statement about a hypothetical idea that has no critical predictions. Moreover, the concept he was referring to, *Many Worlds*, simply cannot be tested empirically, as far as is currently known. It's a classic case of *Nullius in Verba*.

The *Many Worlds* hypothesis is an example of a *vacuous theory*, as I call them. Such "theories" have no empirical evidence in their favour and, more significantly, have no foreseeable prospects of any either. *Science Guy* did not mention those facts, so what he said was misleading.

I don't object to mention of *Many Worlds* as a *hypothesis*, no more or less than I would object to mention of *God*, *Heaven*, *Hell*, *Middle-Earth*, or *Fairyland*, but it's not factual based science *per se*. Its status as a vacuous scientific concept should have been emphasized as a priority. In his privileged position, *Science Guy* should have spoken keeping *Nullius in Verba* in mind. How else can scientifically untrained people distinguish proper science from science fiction?

### *Working in a closed information bubble*

I am a physicist (at least I like to think so), but I don't recall ever being asked my opinion about this particular hypothesis. So regardless of whether the hypothesis is true, I don't see on what basis *Science Guy* felt justified in talking about the "majority of physicists". Did he mean the majority of people he's talked to? If so, then that would be an example of a *closed information bubble*. That's a situation where a person who believes in, say, the *Flat Earth hypothesis* reads only articles and books devoted to proving the Flat Earth hypothesis, and only talks to people who believe in it. Would *you* trust such a person to give you a fair-minded view of the Flat Earth hypothesis? I hope not.

We all live in closed information bubbles. They are often unavoidable. Whether we go to school or work in an office, the people we meet will share many of our interests and opinions, and it's unlikely that those opinions will get seriously challenged in those environments. It's important to be aware of this and take reasonable steps to burst whatever bubbles we find ourselves in. Otherwise, we may get conditioned into modes of thought that are in the long run unhealthy.

The way to avoid closed information bubbles is to read and listen as widely as you can about all sorts of ideas, examine the *empirical* evidence for them, and then work out good reasons why they might be wrong and misleading, or why they are right and everything you believed in before is wrong.

I suspect *Science Guy* operates in a closed information bubble and talks to more theorists who believe in the *Many Worlds* hypothesis than dismiss it as vacuous.

### *Science is not a democracy*

*Science is not a democracy and one person's opinion is not as good as anyone else's.* Science, if it is to be done properly, cannot be run on the basis of a popularity contest, as implied by *Science Guy*. The truths of science are dictated by the laboratory and not by opinion. It should not matter one iota if ninety-nine percent of all scientists, or even if one hundred percent of all scientists, believed in the *Many Worlds* hypothesis, or any other idea for that matter. *Nullius in Verba* tells us that.

## 22.2 Why does it happen?

Why would *Science Guy* say what he said? That's a relevant question and I think I know the answer: *hype sells. Science Guy* happens to write popular science books that are designed to entertain the reader. They sell in many more numbers than this book will ever sell[1]. They won't tell the reader that proper theoretical and empirical science are *hard work, frequently lead to dead ends, and are not part of the entertainment industry.*

Hype is the intensive promotion, publicization, and exaggeration, of some idea or product, designed to condition the listener into think of that idea as true, or that product as desirable. This is based on the well-established fact that the more you hear some idea being repeated, the quicker you will get conditioned to accept it as reasonable. It's an insidious method of conditioning people into accepting the unprovable and getting around *Nullius in Verba*. It's a technique used by many different kinds of organizations.

You can find examples of hype all over the internet. Look at *YouTube* and observe how many contributions are labelled "*the funniest thing you ever saw*", or "*you'll love this!*". Frankly, *Nullius in Verba* tells me that if a title starts like that, then there's a good chance that the video really is *not* the funniest thing I ever saw. *Nullius in Verba* has conditioned me to be alert to the tricks of adverts. Whenever any advert tries to tell me what my opinion should be, I automatically take the opposite view. I can make my own mind up. It's because of *Nullius in Verba* that I knew what was wrong with *Science Guy*'s broadcast that morning.

## 22.3 Heresy

A *heresy* is a statement that is regarded by a group of people as opposite to their conditioned beliefs, and therefore a threat to their values, way of life, and society. There have been many examples of heresies throughout history. Ever heard of *Akhenaten*? He was a Pharaoh in Ancient Egypt who tried to overthrow the religion practised before him with one favoured by himself. Pharaohs who came after him did their best to erase him from the historical records.

Heresies are usually associated with religions, but I like to think that they have a wider applicability to any sort of clash of ideas. It really depends on how ingrained the conditioned ideas are and how radical the new, heretical, idea is in comparison. Heresies can crop up in science. At one time, a belief in the Heliocentric theory (the one that says the Earth goes around the Sun and not the Sun around the Earth) might have led you to a public burning. Nowadays, if you went into some University departments and suggested that *String Theory* or *Many Worlds* are wrong, you might be branded as a heretic. In the case of *String Theory* and *Many Worlds*, the charge of scientific heresy might just, perhaps, maybe, possibly ... more correctly be *pointed the other way.*

Given the hype associated with *String Theory* and *Many Worlds* I'm going to deal with each at some length. You as a potential mathematical physicist may find some of what I write of interest, and, I hope, of concern.

## 22.4 String Theory

Because I set out to alert non-experts about the dangers of ignoring *Nullius in Verba*, I feel obliged to discuss *String Theory*. I regard the discussion as more important than the theory itself, which for reasons I'll explain, is currently vacuous, and has every prospect of remaining vacuous. A significant issue is that *String Theory* has consumed the intellectual, financial, and career resources of countless mathematical physicists for reasons more to do with sociology than *String Theory*'s scientific merits.

I never worked in *String Theory*, a fact for which I am immeasurably grateful. You may well ask how qualified am I to comment on this subject?

I was a research student doing a doctorate in theoretical high energy elementary particle physics when *String Theory* first appeared, in a department that included one of the four original authors of the first paper on *String Theory*. I watched on the sidelines as *String Theory* developed over the years. I saw what happened in 1984. Coincidentally, *Ninety Eighty-Four* is the title of Orwell's frightening book about a society where truth is fabricated and history erased and rewritten [Orwell (1949)].

Before 1984, *String Theory* was a moderately useful heuristic (hand-waving) phenomenological model of strong interactions in high energy particle physics. After 1984, *String Theory* no longer claimed to be phenomenological but a complete theory capable of "explaining" General Relativity and the whole of elementary particle physics. *String Theory* is now a mathematically sophisticated enterprise that has spread its tentacles into various corners of mathematical physics such as cosmology and particle physics, consuming enormous resources and careers with no empirically verified *critical prediction* to date.

I'm not the only person who has serious misgivings about *String Theory*. Some mathematical physicists who have worked in and contributed to that theory think as I do [Woit (2006); Smolin (2006)]. It's important for you as a potential mathematical physicist to understand the history of *String Theory*. Then you'll be able to make a reasonable judgement as to whether you want to contribute to it or avoid it.

**You** may well find yourself faced with such a decision if you go into the world of mathematical physics. Here's a rather sad anecdote about *String Theory* that may shake you up a bit and alert you to the dangers of hype and spin in science. This anecdote concerns students who were sucked into the world of String Theory and came to regret it.

### Quicksands

It was in the middle of a three week-long Summer School in particle physics, in Oxford. I was there in the capacity of an observer, being interested in the field, listening to the talks but not contributing. I was reviewing in my mind my thoughts about the subject, hoping to get some insights into new areas to work in. Keeping in touch is after all a sensible strategy for a mathematical physicist.

One night during the meeting, I found myself in a local pub with a number of young post-doctoral students who were all doing String Theory.

Things were going well, they were a fine bunch of people.

We exchange jokes. Then one of them asked me a riddle.

"*Do you know what do we call ourselves?*" pointing to his colleagues in String Theory.

I frowned. "*String Theorists?* I ventured.

"*No,*" came his reply.

"*We call ourselves ... **String Victims**.*"

As in all such issues, the critics of *String Theory* may be wrong. In fifty year's time, perhaps *String Theory* will have been vindicated. But arguing that *mathematical beauty* counts more than empirical validation, which some *String Theory* theorists advance as a reason for doing *String Theory* (in the absence of a validated critical prediction), is contrary to the principles of science and deserves having *Nullius in Verba* thrown back at them.

## *Background*

As I have indicated earlier in this book, I started a doctorate in high energy elementary particle physics. By the late nineteen sixties, many high energy collision experiments had revealed strange phenomena that were different to anything encountered in ordinary life. In particular, families of particles with uniformly spaced masses were inferred from the shape of scattering cross-section data plots. These families, called *resonances*, played a decisive role in the origin of *String Theory*.

To explain these families, mathematical physicists devised a model based on *quarks* and *gluons*. Quarks can be though of as hard particles whilst gluons are part of the "glue" (hence the name) binding the quarks to each other to form composite objects, the so-called resonances that I refer to above.

There is an analogy here with the forces of electrodynamics that's useful in understanding where *String Theory* came from. You may have heard of *electrons*. These are tiny, electrically charged particles that are constituents of ordinary atoms. You can think of the electrons in an atom as orbiting the nucleus, which is the much heavier, oppositely charged centre of the atom.

Physicists discovered several decades before *String Theory* appeared that there are oppositely charged electrons in this universe. They're called *positrons*. They have exactly the same inertial mass as ordinary electrons and all other things about them appear the same, except for one astonishing fact: they have an equal but opposite electric charge.

According to the laws of electrodynamics, equal and opposite electric charges attract, so an electron and a positron will attract each other to form a composite object called *positronium*. What keeps the electron and positron bound together is the electromagnetic field. That field acts as the glue that forces opposite charges together and, surprisingly, forces like charges to repel. Particle of light, known as *photons* are the electrodynamics equivalent of gluons that bind quarks and anti-quarks together.

It's in the glue concept that reside many of the fundamental technical issues haunting mathematical physicists over the centuries. Take the photon glue, the electromagnetic field. We can literally "see" it, because as the mathematical physicist James Clerk Maxwell showed, light is none other than vibrations in that glue. Those vibrations travel, of course, at the speed of light. Another glue that physicists have known about for centuries is *gravity*. There's a strong suspicion amongst many mathematical physicists that there should be particles of gravitational glue, just like photons are particles of electromagnetic glue. Such gravitational particles are known as *gravitons*. Significantly, gravitons have not been observed to date, and it is an open question in mathematical physics as to whether they should exist. Gravity is remarkably different to electromagnetism, in that the gravitation we are familiar with is always an *attractive* force, whereas electromagnetism can have attractive and repulsive force aspects, depending on the nature of the electric charges.

The experimentalist Michael Faraday popularized a diagrammatical way of comprehending electromagnetism. He imagined electric and magnetic field lines running in space from electric and magnetic charges. In Figure 22.1, diagram (a) shows a sketch of the electric field lines running from a positron (labelled $+$) and an electron (labelled $-$).

In *String Theory*, the picture for the gluon field lines between a quark (labelled $q$) and an anti-quark (labelled $a$) would be similar, except for one critical difference. In quantum field theory terms, gluon fields *self-interact* whereas electromagnetic fields do not. Understanding this difference was a major triumph in mathematical physics. The consequence as far as quarks and anti-quarks are concerned is profound. Figure 22.1 diagram (b) indicates via the big arrows an effective pressure that results in the gluon lines of force being squeezed into a narrow tube of *flux* running from the

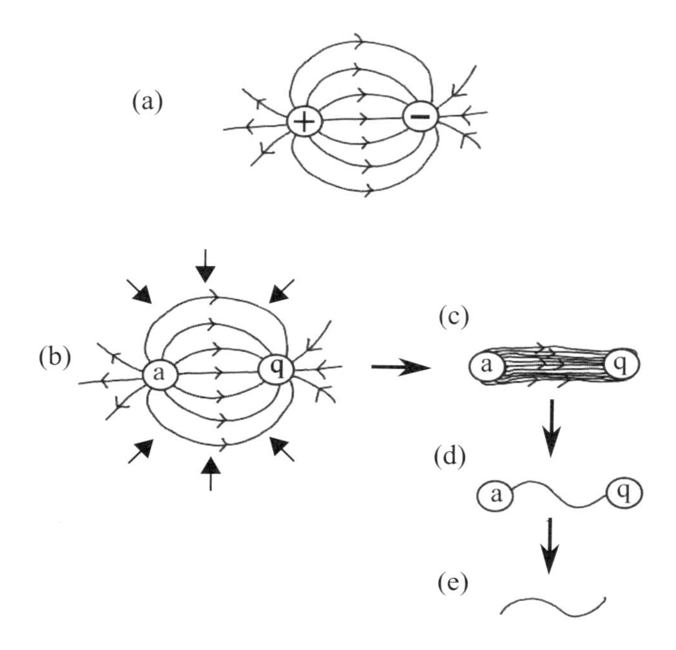


Fig. 22.1 How *String Theory* was born.

anti-quark to the quark (or vice-versa, depending on your convention). This gives Figure 22.1 diagram (c).

The flux tube concept was believed to account for the resonances mentioned above, the families of observed particles with equally spaced masses. Imagine a guitar string. Unplucked, it lies in its lowest vibrational state, which is not vibrating at all. That state was imagined to correspond to empty space, which is an absence of any particle. Now imagine plucking that string. It will have a whole spectrum of vibrational states, corresponding to different frequencies of the sound emitted. Each possible vibrational state has an energy and different vibrational states have different energies. The idea was that the different frequency vibrations of a gluon flux tube would correspond to the different observed particle types. Moreover, since energy and mass are linearly related by Einstein's formula $E = mc^2$, the hadronic string model seemed to give a neat explanation for the resonances observed.

Once theorists suspected such a phenomenon might be in operation, it was a logical step to idealize the flux tube to a thin line, a *string*, running from the anti-quark to the quark, diagram (d). That string is the original string of *String Theory*.

Mathematical physicists tend to idealize approximate models. It was not long before the quark and anti-quark were thrown away (conceptually) and mathematical physicists started discussing *open* strings, which are strings without quarks or anti-quarks on their ends, diagram (e). A logical extension was to join the two ends together to have *closed* strings.

The length scale associated with such *hadronic strings* has a lot to do my discomfort with modern *String Theory*. In the 1950's, it was established that the "size" of the proton, one of the so-called elementary particles, is about $10^{-15}$ metres. It's not an exact number because size here depends on what you're measuring. Whatever method you use, the proton is about five thousand times smaller than a typical atom, which is about a million times smaller than a human hair. The length of a gluon flux tube would be similar to the size of a proton, so the original strings as discussed here were thought to be on the order $10^{-15}$ metres long. It's not a critical problem here if I'm out by a factor of ten or a hundred either way. What happened next is my concern.

## *That shrinking feeling*

As the years rolled on, the hadronic string model discussed above proved somewhat disappointing. As with any model, it had its good points, but a few major blemishes as well. One of these was *quantization*. The world of theoretical particle physics is structured on quantum mechanics, not classical mechanics. It's true that classical mechanics is used as a starting point, guiding the mathematical physicist in formulating equations of motion, and suchlike. However, sooner or later, quantum mechanics is needed for high energy physics.

Given a classical model, such as the hadronic string discussed above, mathematical physicists will invariably attempt to "quantize" that model. That means re-interpreting the classical variables as quantum operators (you don't need to know what that means here), and a deal of technical details to do with mathematical structures and mathematical consistency.

When mathematical physicists attempted to quantize the hadronic string, the results were surprising. The original model was consistent only in a spacetime of twenty-six dimensions, not the four dimensions (one time and three space) that we think we live in. Moreover, the lowest, unplucked state of the hadronic string, the one corresponding to empty space, looked like a *tachyon*, or a particle with imaginary mass.

Over the years, various amendments were made to the original hadronic string model. It eventually morphed into a class of model known as *superstrings*, for technical reasons that cannot be explained and in any case, are not relevant to my concerns. The consistency dimension of spacetime was brought down to ten in those models, and the tachyonic state eliminated from the formalism. There were many discussions as to what the extra dimensions in the string corresponded to.

Recall that the length scale of the original hadronic string was $10^{-15}$ metres. The superstrings that are discussed by String Theorists these days are posited to be on the scale of $10^{-35}$ metres, a length scale known as the *Planck* length. It is believed by String theorists to be the shortest meaningful distance that should be discussed in physics.

My problem is not so much with *String Theory* as a mathematical theory. My problem is with the mathematical physicists who changed the scale of their strings spontaneously:

> A whole cohort of string theorists went to bed one night thinking of strings as being about the size of a proton. When they woke up next morning, they starting talking about strings that were a hundred million million million times smaller than they were the day before.

That's what happened in 1984, during the so-called *First String Revolution*. It reminds me of the Klingons in Star Trek, who suffered some unexplained genetic mutation that transformed them into totally different looking creatures as they passed from the Original Series to the Next Generation.

No one blinked and said "*Excuse me. Could you explain exactly what you've just done?*"

The mathematical technologies that had been developed for hadronic strings were completely transferable to this new, fundamental string theory. That meant that the generation of mathematical physicists who had worked on hadronic strings were not suddenly unemployed.

Since 1984 there have been many technical developments in *String Theory* some of which looked exciting at first sight. For example, there was a hope that gravity could be brought into the picture, because *gravitons*, the hypothetical quanta of gravitation, are naturally associated with closed strings. Developments in *String Theory* reveal tantalizing glimpses of deeper relations between standard particle theory, General Relativity, and cosmology. For that reason, many mathematical physicists continue to work in *String Theory*.

The problem that has been pointed out by critics is that so far, no new observable physics has emerged out of *String Theory*. The question is: does *String Theory* go the way of the epicycle theory of planetary motion, or of the atomic structure of matter hypothesis?

## 22.5 Many Worlds

I turn now to my review of the *Many Worlds* paradigm. Given a choice between *String Theory* and *Many Worlds* I would take *String Theory* any day. It may be simply a matter of time before some really high energy experiments reveals data that *String Theory* is required for. I think there's no prospect of anything like that as far as *Many Worlds* is concerned, because it's too close to metaphysics, in my view.

*Many Worlds* came about because quantum mechanics is different in many technical and conceptual ways to classical mechanics. I could go on at length about these differences, but that topic deserves a complete book on its own. There's one aspect of quantum mechanics that may help in explaining the issue here and that's *quantum superposition*. To discuss that, I need to go on a bit about the underlying principles of classical mechanics and then of quantum mechanics.

### *Classical principles*

In classical mechanics, suppose a particle could be in one or other of two possible places, and no other. Let's call those places $A$ and $B$, and this situation the $AB$ experiment. In classical mechanics, we would describe the state of the particle being at $A$ with a mathematical representation of some kind. Let's denote that representation by the symbol $C_A$, where the "C" stands for "classical". Likewise, if the particle was at $B$, then we would denote that classical state by $C_B$.

There is a fundamental principle in classical mechanics. According to classical mechanics, the classical state of the particle is *either* $C_A$ *or else* it is $C_B$. We would *never* allow it to be anything other than a strict choice between the two states. That's because of the classical principle that a single particle cannot be in two places at the same time.

Another fundamental principle in classical mechanics is that we need not actually look to see where the particle actually is. It is a cardinal tenet in classical mechanics known as *realism* that the particle "exists" in just one of the two possibilities, $A$ or $B$, for sure at any given time.

I have a few comments about this.

**1.** There's that word again: "*exists*". If you've read my anecdote *It Exists* in Chapter 19, *Sitting at the Back*, you'll now appreciate that the philosopher I had an issue with was obviously a realist, even if he did not say it or know it. It was his misfortune that I was beginning to have my serious concerns with realism when I heard his talk.

**2.** Realism is a conditioned view of our environment based on what we think we see with our eyes. We are deceived by the vastly complex processes of vision that end up creating a mental map of whatever is in sight, and this map conforms to the principles of classical mechanics. It's a convenient mental trick played by our brain on itself to overlook this fact and convince ourselves that this mental map *is* reality. Close your eyes and the map disappears.

**3.** I'm not saying that reality is an illusion. I'm not saying that we see illusions. I'm saying that what we think we see, those thoughts are the illusions. Any neuroscientist will tell you that it's our brains that create illusions from sensory information that comes in from our environment. Data by itself has no quality of illusion or truth. Illusions are, after all, no more and no less than a misidentification of data. For example, a mirage is no longer an illusion once we have understood that we are seeing distortions of light optical paths via atmospheric disturbances. Another example is *magic*. Magicians manipulate ordinary matter in such a way as to make their audiences believe it's magic. Magic and illusions do not "exist" except in the mind.

The realist interpretation is manifestly incorrect in principle for the following reason. If we're scientists, we simply cannot "prove" any assertion about reality, whatever that is, without empirical evidence. But we cannot have empirical evidence without observation. We cannot simply wish away that observational side of it. It's always there. Only a philosopher goes a step too far and invites us to ignore observers, apparatus, and observation (usually quoting some fancy classical principle such as *general covariance*). Any arguments like that invariably fall foul of *Nullius in Verba*.

**4.** The realist interpretation is extremely effective usually, but breaks down when detector technology is good enough to see its flaws. It was advances in technology that allowed scientists to see those flaws and led them to devise quantum mechanics in order to explain their data.

However, mathematical physicists are humans, with all the classical conditioning of humans. That leads a sizeable fraction of mathematical physicists to cling to realist concepts whenever they can, even when they don't realize they are doing so. Even Einstein clung to the realist view of reality, which is why he ended up disliking quantum mechanics.

**5.** I don't say reality does not exist outside of observation. It would be contrary to *Nullius in Verba* to make any assertion I could not empirically validate. Therefore, I say nothing about things I cannot deal with empirically. *Saying nothing is really a hard thing to do, usually.*

**6.** I have noticed that all realist models such as *Hidden Variables* and *Many Worlds*[2] invariably use language that implies some exophysical observer "looking in" on proceedings. Be wary of such language, because that's where tricks are being played on the unsuspecting listener.

### *Quantum principles*

From the year 1900 onwards, when Planck published his papers on quanta of energy, it was found that the above classical mechanics scenario was not good enough to explain all data. In a quantum mechanics description of the $AB$ experiment discussed above, the particle being at $A$ would now be represented by a quantum state denoted $Q_A$, where "Q" stands for "quantum". Likewise, the particle being at $B$ would be describe by quantum state $Q_B$.

Here's the really big deal about quantum mechanics. It was found that time plays a big role in all of this. In the classical discussion, I didn't say anything about who was preparing the state of the particle or when it was going to be determined (by looking at $A$ and $B$ for the particle). It turns out that you've got to be much more careful than that. Now *context* plays a huge role.

*Context* is my word for taking everything relevant into a physics discussion. My view is that many if not all the conceptual issues people have had with quantum mechanics over the last century have been because some context has been ignored in the discussion. Indeed, the realist position goes the whole way and says that particles and their physical properties "exist" regardless of observational context.

In the case of the $AB$ experiment, for instance, we should ask a realist *how do they know that there is a particle there in the first place?* That's not a small point.

I said that time plays a crucial role in all this. Why is that? If you accept the need to tell me how the particle state was prepared in the first place, then you are automatically introducing time into the discussion. Preparation is, after all, something over and done with by a certain time (otherwise, you never end up with a prepared state). You would be invoking time also when you observe where the particle was, *after* state preparation (obviously not before).

Consider the time interval after state preparation and before observation. It was discovered that, according to quantum mechanics, you simply could not assert that the particle state had to be either $Q_A$ or else $Q_B$ during that time interval. Indeed, depending on how you prepared the state of the particle, you might have to have a *superposition* of the two states $Q_A$ and $Q_B$, of the form

$$\alpha Q_A + \beta Q_B. \tag{22.1}$$

Here I am going a bit deeper into the mathematical formalism than I would like to in this book, and am actually employing the mathematics of *Hilbert spaces*, which allows me to use addition of states as above. The constants $\alpha$ and $\beta$ are numbers that depend on how we had prepared the state of the particle[3].

In essence, *before* we looked, we could not assert as we did in the classical case that the state of the particle was either $Q_A$ or else $Q_B$.

Now imagine we looked and actually found the particle at $A$. We would no longer be entitled to say the particle's state was given by superposition

(22.1), but by $Q_A$. In other words, performing an observation would change our state description of the particle.

It here that many of the conceptual issues people have with quantum mechanics come from:

### *Superposition*

Many people do not like the idea that a particle could be represented by a superposition such as (22.1) of two otherwise "normal" states. That dislike led eventually to the idea that there are multiple universes, such that, in each, a particle is always in a "normal" state. Followers of this *Many Worlds* idea tend to be rather confusing as far as I am concerned as to what precisely they believe. There are several versions of *Many Worlds*, all of which fail to explain this or that concern of mine.

Two of these concerns are:

**1.** The wavefunction or overall state of the Multi-Universe (the totality of all Universes in a *Many Worlds* model) is taken to "exist", with no reference to any external observer. That means there is an appeal to some form of realism. That clashes directly with *Nullius in Verba* as far as I'm concerned.

**2.** Despite assertions to the contrary, *Many Worlds* models do not "explain" the probabilistic rules of standard quantum mechanics. In analogous experiments, it was found that if you ran the $AB$ experiment a billion times, the number of times you found the particle at $A$ would be proportional to $|\alpha|^2$, and likewise, the number of times it was found at $B$ would be proportional to $|\beta|^2$. No one really "understands" why in a convincing way. That's one of the great unsolved problems in mathematical physics. Maybe you will solve it.

### *Wavefunction collapse*

Many people believe that the change from superposition (22.1) to $Q_A$ as a result of observation represents an actual collapse of something objective. In fact, it is just a change in the *description* that you the observer makes in your mind as to the state of the particle. As you get more information, so you adjust your description. That's no different to the "collapse" of probability that you get once the result of a football match is announced. No one actually imagines that the chances of a team winning a football match is an objective thing. Likewise, a quantum state is a mathematical representation in the mind of a theorist. Once new information comes in, that state will change.

## 22.6 Hidden Variables

The only thing I need to do here is to throw *Nullius in Verba* at the whole idea of *Hidden Variables*. Usually, I find papers on *Hidden Variables* leave out just sufficient context for their discussions to appear reasonable. Look at *Hidden Variables* concepts carefully enough and the deficiencies in *Hidden Variables* become obvious.

## 22.7 Commentary

I would not have any objections whatsoever to *String Theory* or to *Many Worlds* (in any of its possible forms or interpretations), if there was some critical empirical evidence for them at this time (2019). Currently, there is none.

Although I am wary of these ideas, I am mindful of the so-called "experts" who dismissed space travel as impossible, a few years before Yuri Gagarin became the first human in outer space in 1961. It's too easy to be a negative critic.

On the other hand, I'm familiar with various failed scientific *boondoggles* such as Aristotelian mechanics, the Ptolemaic theory of epicycles, the caloric theory of heat, and the continuum theory of matter. Mathematical physics has had its particular share of *boondoggles*, such as the *Fourier principle of similitude* (the laws of physics on the smallest scales are the same as those on the largest scales), the Aether (its vibrations are light waves), and a variety of "bandwaggons" in elementary particle physics too numerous to list. Such bandwaggons arise during periods of uncertainty in physics.

To my mind, the problem with all of these topics stems from a failure to encode observers and the processes of observation properly into the formalism. I don't think it can be done in *Hidden Variables* at all, by definition. I think *Many Worlds* might do it, but would be destroyed, as it stands, in the process. I think *String Theory* could do with it and perhaps survive unscathed. Perhaps *you* might have a go.

# Chapter 23

# Notation

*An important step in solving a problem is to choose the notation. It should be done carefully. The time we spend now on choosing the notation carefully may be repaid by the time we save later by avoiding hesitation and confusion. Moreover, choosing the notation carefully, we have to think sharply of the elements of the problem which must be denoted. Thus, choosing a suitable notation may contribute essentially to understanding the problem.*

[Polya (1971)]

In Chapter 10 (*No Good at Maths?*), I referred to *symbols* in mathematics. I said there that symbols are marks on paper that convey a message. Actually, they are much more than that. In the language and philosophy of Sheldon Cooper of *The Big Bang Theory* fame, symbols are part of a *non-verbal, unwritten contract*, a social convention if you will, between two sorts of people who may perhaps have never met, or even have had no possibility of ever meeting. Let's call them *senders* and *receivers*. A *sender* makes marks on paper or other media such as clay tablets, stone, or electronic files. Later (and it is always *later*), a *receiver* (or in the case of widely distributed documents such as newspapers, books, and the internet, *receivers*) looks at those marks and tries to make sense of them. Those marks are the symbols that I'm discussing here.

There are several important points about symbols that you should understand if, say, you are put off mathematics because it "looks hard".

## 23.1 Visual impact

If a densely printed book of mathematical symbols turns you off the first time you look at it, welcome to the club. You would not be alone. For instance, if you saw for the first time one of the pages of a book I bought recently containing *Srinivasa Ramanujan*'s mathematical papers [Ramanujan (2000)], I think you would have my reaction when I first got that book: *no way could I understand that*[1]. I bought that book as if I were an art collector acquiring a *van Gogh picture*: it's something to cherish and take pride of ownership in having. Besides that, there's always a chance a visitor to my house might believe I could understand any of Ramanujan's work.

When I first started looking at mathematical textbooks at school and later at university, I usually made the simple mistake of thinking that I had to understand them first time around. The result invariably was that I would read an introduction and then generally come to a depressing halt some time after page ten, discouraged by my apparent lack of ability in mathematics.

It's important not to be taken in by such self-doubts. Would you have the same reaction to learning to play a piano, or learning to drive a car? No. In such cases, you would start at the beginning, take easy lessons at first, and gradually work your way up into more technical aspects. So it is with mathematically based textbooks. You have to treat them as if they were potentially dangerous devices.

I'm not exaggerating. A book that looks impossible to understand could make you believe that it's **your fault**; that **you** are the problem and not the book. That may well be the case if you're a complete beginner, but there's always the possibility that the book was badly written, or aimed at experts with years of training behind them.

## 23.2 Nothing is perfect

A *typo* is a common diminutive for *typographical error*, or misprint. Typos are the viruses of all written documents, either handwritten or typeset. Everybody makes typos.

I've made many typos in my handouts to students, but never by design. Students should be alert to typos as they can cause serious loss of confidence.

**Typos**

Just imagine you're an enthusiastic beginner in some mathematical subject. You've bought an expensive book and you start to work your way through it, line by line, page by page. You do all the calculations successfully until ... you run into a road block. A formula is given in the text and you try to reproduce that formula with your newly found knowledge. But something doesn't quite work out. Repeatedly. You try and try for hours, days, and weeks, but no, you can't get that factor $c^2$. Every time, you come up with $c^3$ instead.

It gets serious. Now you start to believe that it's *you* that's at fault. Why can't you get that factor $c^2$? Surely you've done something stupid. So you try again, and again.

Finally, you've had it. Obviously you're not as good a theorist as you thought you were. In frustration, you throw the book to the floor, determined to quit mathematical physics.

And it's only then, as the book tumbles to the floor, that the *Errata* page comes to light. It's a loose page inserted at the back of the book by the publishers of that book, informing readers that there's a typo in formula 3.12 on page 18, where a factor of $c^2$ should be $c^3$.

I have a book on quantum field theory that has 705 pages of text [Itzykson and Zuber (2005)]. The edition I bought has a printed chapter entitled *ERRATA* before the text starts. That chapter consists of ten pages, containing a total of 120 itemized errata, two of which are a page long each. Imagine trying to learn the subject without that ERRATA chapter.

Perhaps it would have been kinder just to publish the ERRATA on its own.

I wonder if there are any typos in that ERRATA?

## 23.3 Take it easy

In this section I'm highlighting an important issue that I think has affected many people. Suppose you had gotten over the issue of visual impact that I mention above, but were attempting to read a book that was beyond your level of expertise *at that particular time.* If you don't make progress reading into that the book, there's the possibility that you might be put off that subject, perhaps for life.

With this in mind, you should tread carefully with mathematically based books. *Read them carefully and be kind to yourself.* You cannot be expected to absorb every subject quickly. Sometimes you will be in for an intellectual beating. A good subject deserves you spending a lot of time on it. There will be subjects that may take *years* before you are conditioned mentally to think about them in a natural, relaxed way. All that is needed is a careful, systematic, patient approach, confidence in yourself, lots of time, and lots of practice.

## 23.4 Consensus

It can be a mistake to get one book and stick to it. What if the maths or physics book you bought is just badly written? What if it's out of date in its notation and subject matter? It's even possible that the author is simply wrong in their statements or core beliefs about the subject they're writing about. You have to watch out for that. Some authors may well have an agenda, such as trying to disprove Relativity or quantum mechanics. It does happen. I know, because I have some books that were written by such people. I bought them before I knew that such games were played, even in academic circles.

Whenever I started a lecture course, I would be expected to give some recommended text that the students could use to supplement the lectures. I always thought that was a *bad* idea. There are no perfect textbooks, but some are excellent. In my view, the best policy is to select your particular concept and consult several books on what the authors have to say about it. There's nothing finer than to walk around the shelves of a library, picking up books more or less at random and scouring their indexes for a particular topic. You soon get the feel of the subject and, more importantly, you'll discover that very few authors agree totally on their conventions, notation, and, sadly, content of their books. I would try to get an overall consensus, taking as many notes as I could on the topic I was interested in currently.

## 23.5 Agreement

It's no good if the symbols used by a sender don't mean anything to a receiver. Both parties have to believe that the symbols being used mean the same. When there is no agreement, as in the case of a foreign language that you haven't learnt, the symbols are meaningless (but only to you).

Here's what happened to me one lunchtime.

### The Cubic

It was lunchtime in the maths department where I worked. I was alone in my office eating a sandwich when the phone rang. It was one of our secretaries down in the main office. Someone from Engineering had rung up seeking help with an equation. No one else in Maths seemed to be around, so could I advise?

In my experience, it's not good policy to say *no* until you know what you're dealing with. So reluctantly, I agreed to speak to the Engineer, and the secretary duly connected him through to me.

The Engineer's disembodied voice introduced himself and asked if I knew how to solve a *cubic equation*. His research team had come across a mechanical problem that required them to find the solution to a cubic equation. The trouble is, none of them knew how to solve such a beast.

These days, of course, you'd just go on the internet and get the formula for solving a cubic equation. But in those days, there was no internet. Unbelievable, but true.

As it happened, I knew two things about cubic equations. First of all, they are *very hard to solve* on your own. In fact, the history of Mathematics records centuries of effort by great minds to solve cubic equations. The ancient Babylonians, Egyptians, Greeks, Indians, and Chinese, all had come across this beast. It was solved mainly by Italians in the sixteenth century CE, so I was not surprised that the Engineers couldn't solve it.

The second thing I knew was that I had a book with the solution.

You've probably been taught how to solve *quadratic equations*. These are of the standard form

$$ax^2 + bx + c = 0,$$

given that $a$ is not zero (otherwise the equation is no longer a quadratic one). The problem is to determine what values of $x$ satisfy this equation if $a$, $b$ and $c$ are constants.

The two possible solutions are given by the standard formula

$$x = \frac{-b \pm \sqrt{b^2 - 4ac}}{2a}, \quad a \neq 0.$$

It's easy enough to memorize.

After discussing their problem over the phone for a few moments, the Engineer and I eventually agreed that he needed to solve an equation of the form

$$ax^3 + bx^2 + cx + d = 0,$$

where now $a$, $b$, $c$, and $d$, are constants, with $a$ not equal to zero (otherwise the equation reduces to a quadratic, which everyone knows how to solve).

As I said, my secret weapon was that I knew that the book *Formulas, Facts and Constants* by Fischbeck and Fischbeck contained the formula for a cubic [Fischbeck and Fischbeck (1982)]. I didn't let on I had this book. After all, you don't give away *all* your trade secrets. That and the fact I wanted to look smart.

That book has helped me out countless times. As a mathematical physicist, you'll soon find that collecting reference books, tables of integrals and special functions, and obscure formulae becomes part and parcel of the game. A mathematical physicist needs their tools of the trade just as much as a plumber needs their spanners and whatnots. Even when you're young, you soon find you can't remember obscure formulae accurately, and there are many situations where a single error in a mathematical sign (plus or minus) can wreck your otherwise great theory. So you start to create your own collection of useful facts and theorems. Mine are several decades old and still of great value to me.

I grabbed the book and opened it on page 34. There it was! The great formula for the solutions to the cubic equation. And I do mean *great*. It's not simple. Here's what Fischbeck and Fischbeck give:

*Given* $ax^3 + bx^2 + cx + d = 0$, *first rewrite it in the form*

$$y^3 + py + q = 0,$$

*where*

$$y \equiv \frac{b}{3a} + x,$$
$$p \equiv \frac{c}{a} - \frac{b^2}{3a^2},$$
$$q \equiv \frac{2}{27}\left(\frac{b}{a}\right)^3 - \frac{bc}{3a^2} + \frac{d}{a}.$$

*Then the three solutions* $y_1, y_2$, and $y_3$ *are given by*

$$y_1 = u + v,$$
$$y_2 = -\frac{1}{2}(u+v) + \frac{i\sqrt{3}}{2}(u-v),$$
$$y_3 = -\frac{1}{2}(u+v) - \frac{i\sqrt{3}}{2}(u-v),$$

*where*

$$u \equiv \left(-\frac{q}{2} + \sqrt{\frac{q^2}{4} + \frac{p^3}{27}}\right)^{\frac{1}{3}},$$
$$v \equiv \left(-\frac{q}{2} - \sqrt{\frac{q^2}{4} + \frac{p^3}{27}}\right)^{\frac{1}{3}}.$$

Now imagine you were on the phone, trying to read all of these details to someone on the other end. Remember, there was no Skype, Facetime, or other video call means for me to show the formula. I had to talk it through.

Quite a long time later, we finished. We had gone over the formula several times, so as to get it right. So then I said "*That's it. Any questions?*"

There was a long pause, and then the Engineer asked...

"*What's i?*"

...

Look at the above expressions for $y_2$ and $y_3$. The symbol $i$ appears next to the square root of three. It's the symbol mathematicians, physicists, and mathematical physicists, use for the square root of minus one.

I was stunned. Surely the Engineer would know all about that. You've possibly done it at school. I pressed the point. "*It's the square root of minus one, of course!*"

There was a long pause. Then suddenly all was clear. "*We use $j$!*"

And that is exactly right. Engineers often reserve the symbol $i$ for *electrical current*, so to avoid confusion they've settled for $j$ to represent the square root of minus one. But if you didn't know that, you would be lost.

**Learning symbols is hard work**

The human brain is constructed to be flexible and reprogrammable, in the sense that it can be conditioned to accommodate almost any sort of belief structure. *Flat Earth, anyone?* But conditioning can't be done overnight normally. Habits of thought take time to be acquired. That's why children take decades to grow up and be educated. Most other animals don't have such a reprogrammable capacity. They have instincts that are hard-wired into their brains, so they "grow up" relatively quickly.

Once you see this, then you will understand why maths may seem difficult. It takes a lot of time to learn maths, but it is doable. All it takes is time, patience, time, good teachers, time, and more time. And of course, practice, practice, practice. I cannot guarantee that you will be another *Ramanujan*, no more than I can guarantee that you will become another Einstein. But if you practice your maths, you'll get better and better. First learn the symbols and what they represent. Then start to manipulate them into the patterns that mathematicians deal with. It's not impossible that perhaps *you* will turn out to be the next *Ramanujan*.

## 23.6 Notation

The choice of symbols and how they work together, which is what I'll call *notation*, is a really big deal. Arguably notation is the biggest deal in mathematics. If you use the "right" notation, many problems in mathematics and mathematical physics can be done quickly and elegantly.

I'll give you an example. It's a "proof" of Pythagoras' theorem, which says that if $a$, $b$, and $c$ are the lengths of the sides of a right-angled triangle as shown in Figure 23.1, then we can say $c^2 = a^2 + b^2$.

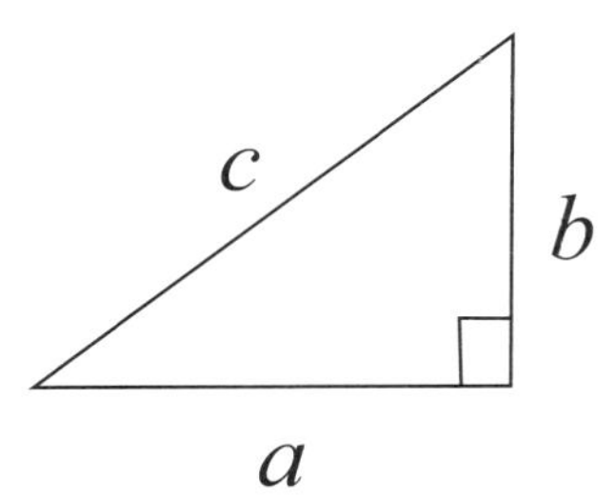


Fig. 23.1 Right-angled triangle with sides $a$, $b$, and $c$.

Now this "theorem" is an ancient one. It was known to the ancient Mesopotamians, Indians, and Chinese. The reason it's commonly named after the Ancient Greek Pythagoras is that the idea that this result holds for *all* right-angled triangles is commonly attributed to him[2]. The usual geometrical proof of this result is usually taught early on in secondary/high school, so you may have come across it. It involves a lot of words and appeals to similar triangles and so on. I won't show it here because it's relatively tedious.

There are other proofs, if you know the relevant notation. Here's a two-line proof using so-called vectors:

$$\begin{aligned}(\mathbf{c} \equiv \mathbf{a} + \mathbf{b}) \;\; \& \;\; (\mathbf{a} \cdot \mathbf{b} = 0) \;\Rightarrow \underbrace{\mathbf{c} \cdot \mathbf{c}}_{c^2} &= (\mathbf{a} + \mathbf{b}) \cdot (\mathbf{a} + \mathbf{b}) \\ &= \underbrace{\mathbf{a} \cdot \mathbf{a}}_{a^2} + \underbrace{\mathbf{b} \cdot \mathbf{b}}_{b^2} + 2\underbrace{\mathbf{a} \cdot \mathbf{b}}_{0}.\end{aligned}$$

### *The summation convention*

Mathematical physicists love to use shortcut notation and often invent their own. There is a convention called the *summation convention* that's often attributed to Einstein and used just about everywhere in mathematical physics. I don't know if he was the person who devised it, but who ever did it was a genius. The convention gets rid of the big summation symbol $\sum$ that we've encountered in Chapter 10, *No Good at Maths?* If you don't use the summation convention, calculations in General Relativity frequently look awful. Use the summation convention, and the same calculations look elegant and you can manipulate the equations readily.

## 23.7 Attraction and repulsion

There is a curious and important side to notation. It can make a subject *attractive* or *repulsive*. It's similar to typeface and font in print. I remember the first time I bought a physics textbook printed in America. It was a revelation, I don't mind saying. Compared to many otherwise excellent textbooks produced in other countries in those days there was a lot of attention paid to visual impact.

A book is principally a visual machine, a device for you to look at. The fact that it's relatively inert doesn't make it any less of a machine than say a computer monitor. It has to function properly in order to be useful. You would adjust the contrast and brilliance of a computer monitor to suit your eyes, wouldn't you? So it is with books. If text on a page is small and hard to read, or lines too cramped together, or diagrams too small to see clearly, then the book can fail to keep the reader's attention.

I found that American college texts tended to be excellent in that respect. Many that I bought (and still possess) had wide pages, with huge margins, beautifully clear text, and wonderful diagrams. These latter were not just your typical line diagrams, either, but often beautifully coloured and shaded pictures that gave a feeling of depth. Reading such a book is an immense pleasure, not a chore.

It should always be kept in mind that mathematical physics involves *communication*, as well as theorizing. Good notation will win a battle for you, but bad notation may lose you the war. Here's a historical example that is relevant to this day.

### Dots and Dees

It could be said with justification that mathematical physics really took off with the invention/creation of *calculus* by Newton and Leibniz.

There are two sides to calculus. There's *differentiation* and there's *integration*. My point here concerns differentiation, a process that turns one function into another. If we are dealing with, say, a function $f(t)$ of time, then we can usually differentiate that function once, twice, three times, and so on, with respect to time (as they say). In fact, almost all the great equations of mathematical physics involve these *derivatives*. Newton's laws of motion all the way to Einstein's theory of General Relativity, they all involve derivatives.

So how do we represent the derivatives of the function $f(t)$?

Newton's answer was to use *dots*. Given a function $f$ of time, the *first* derivative of that function was denoted by Newton as $\dot{f}$ and called a *fluxion* [Newton (1736)]. Note the single dot above the letter $f$. The *second* derivative is denoted by two dots, that is, by $\ddot{f}$ and so on.

In sharp contrast, Leibniz denoted the first derivative by $\frac{df}{dt}$, the second derivative by $\frac{d^2 f}{dt^2}$, and so on.

So which one wins out?

The answer is amazing. For some as yet unknown deep reason, a great deal of mathematical physics involves the number *two*. A good example is Pythagoras' theorem discussed above. It reads $c^2 = a^2 + b^2$ exactly, not $c^3 = a^3 + b^3$ or $c^{2.001} = a^{2.001} + b^{2.001}$.

Technically, a mathematician will tell you that "*the exponent of* 2 *is because of the assumed Euclidean metric structure of space*", but that "explanation" should *not* satisfy a mathematical physicist, in my opinion, because there's a lot of indirect evidence for the notion that space is a physical phenomenon, with its own properties, such as curvature, and not a mathematical construct. We use the Euclidean metric to *model* the geometry of space, but don't necessarily believe that space "is" as simple as that. Indeed, mathematical physicists have played around with generalizations and variants of the so-called Euclidean distance rule, but *two* seems to be fit our Universe best. Traces of this 2 reappear everywhere in science. No one really knows why this is the case. Perhaps *you* will be the one to find out why one day.

Most significantly for our purposes here, Newton's laws of motion (and many others) involve *second* derivatives and no higher, another example of *two* coming into the picture. In consequence, surprisingly many equations in mathematical physics can be neatly written down using Newton's dot notation.

That does not rule out Leibniz' notation, but believe me, if you are working through a ten page calculation, by hand, with hundreds of second time derivatives, Newton wins out over Leibniz.

On the other hand, sometimes we have to deal with large numbers of variables or very large order derivatives, and then the Leibniz notation wins out. For example, the *fifteenth* derivative of a function $f(t)$ makes the Leibniz notation the only one to use. In that particular case we would have

$$\text{(Newton)} \quad \rightarrow \quad \overset{...............}{f} \quad \equiv \frac{d^{15} f}{dt^{15}} \quad \leftarrow \quad \text{(Leibniz)}.$$

It's obvious Leibniz wins in such a situation. Just try to use Newton's notion for the hundredth derivative and you will be convinced.

# Chapter 24

# Your Career Choice

*...as we know, there are known knowns; there are things we know we know. We also know there are known unknowns; that is to say we know there are some things we do not know. But there are also unknown unknowns – the ones we don't know we don't know.*

Donald Rumsfeld [Rumsfeld (2002)]

If you've reached this point in my book, it's possible you will be thinking about which sort of mathematical physicist you would like to be. That's a critically important point to think about, as it may determine the course of the rest of your life.

I have referred at some length to some branches of mathematical physics that I would not currently touch with a barge pole. I may be wrong in my judgement, because if there's one thing that's certain, it's that no one knows the future. *String Theory* and *Many Worlds may* yet turn out to be paradigms that survive, but on present form, I seriously doubt it.

There were periods in the annals of science when theorists complacently believed that they had a proper theoretical understanding of the universe they were observing. One such period ended in the period 1900-1905, when the classical Newtonian view of mechanics became supplanted by Special Relativity and quantum mechanics.

What emerged by the end of the Twentieth Century was an extraordinary state of affairs that continues to this time (2019). By 1915, Special Relativity had morphed into General Relativity, Einstein's greatest triumph. By 1930, quantum mechanics had branched out into numerous successful strands, such as Schrödinger's wave mechanics and quantum field theory.

Ever since, mathematical physics has been dominated by the continued successes of General Relativity and quantum mechanics in their respective domains of applicability.

Despite the spectacular successes of both General Relativity and quantum mechanics in their respective domains of applicability, there remain various fundamental questions that continue to preoccupy mathematical physicists. I discuss a number of these questions now. If you're interested in a career in mathematical physicist, you might well find yourself researching in one or more of these areas for the rest of your life.

## 24.1 Peaceful Co-existence

Wherever mathematical physicists have used General Relativity or quantum mechanics to predict the outcomes of relevant experiments, they have found no apparent conflict between those predictions and the observed data. What worries some theorists is that each of the two strands of mathematical physics, General Relativity and quantum mechanics, appear to be indifferent to the existence of the other. There appears to be no relevant experiment where any prediction of General Relativity clashes with the predictions of quantum mechanics. It's as if General Relativity and quantum mechanics are mathematical theories of non-intersecting sectors of the Universe.

That state of affairs has been labelled *peaceful co-existence* by some theorists who worry about such matters [Licata (2016)].

If you detect some disdain on my part, you would be quite wrong. My working life is fixated on this issue. My personal view of this issue is that there *is* a factor common to General Relativity and quantum mechanics that links the two, but that factor has been so far relatively ignored by both sides. I'm referring to the role of the *observer* and of *observation*. Certainly, both General Relativity and quantum mechanics talk about observers, but they do so from conventional "Newtonian" perspectives. My best guess is that mathematical physicists should eventually be able to unify General Relativity and quantum mechanics into a common framework, but only if this observer issue is developed properly.

What are the problems with doing that? Well, there are already running throughout mathematical physics several conceptual divides that have been recognized and identified by some theorists as having something to do with all this. The problem is, no one currently knows what to do with any of them. I think they all have something essential to do with the Peaceful Coexistence problem.

### *The Heisenberg cut*

One notorious conceptual divide, often referred to as the *Heisenberg cut*, is the imagined divide between the world of classical mechanics, where observers live, and the world of quantum mechanics, where quantum systems live (or so it's imagined). According to Heisenberg, the mathematical physicist who constructed matrix mechanics (a conceptually deep approach to quantum mechanics):

*The dividing line between the system to be observed and the measuring apparatus is immediately defined by the nature of the problem but it obviously signifies no discontinuity of the physical process. For this reason there must, within certain limits, exist complete freedom in choosing the position of the dividing line.*

W. Heisenberg [Heisenberg (1952)]

### *The endo-exo divide*

The *endo-exo* divide is readily explained. Exophysics is a model of observation that supposes that observers look *in* on their experiments, as if those observers were external to whatever was going on. A good analogy is with a conventional theatre audience that simply watches whatever is on stage. On the other hand, endophysics recognizes that observers are *inside* the universe they are observing, so are really part of the process that they are observing. A good analogy is with a modern theatre performance where the players come off stage and mingle and interact with the audience.

When Isaac Newton wrote his great book, The Principia, he based his work on his concepts of Absolute Time and Absolute Space. He wrote:

*Absolute space, in its own nature, without relation to anything external, remains always similar and immovable ... Absolute, true and mathematical time, of itself, and from its own nature, flows equably without relation to anything external, and by another name is called duration.*

[Newton (1687)]

These are exophysical concepts generally used in all conventional formulations of classical mechanics.

### *The reductionist-emergent divide*

*Reductionism* is the idea that simple rules can account for all phenomena, including the Universe. On the other hand, *emergence*, sometimes also referred to as *complexity*, is the idea that the behaviour of systems with very large numbers of mechanical degrees of freedom cannot be described by reductionist principles alone. Often cited examples of emergent behaviour are processes such as life, the laws of thermodynamics, and the origin of the Universe.

Some mathematical physicists subscribe to the reductionist view that there might be a *Theory of Everything.* There's nothing inherently wrong with that idea, because it motivates research in a number of ways, including particle physics and *String Theory.* My personal view is that it does not take into account the fact that both General Relativity and quantum mechanics are *not* actually reductionist theories, although in many places they look like, and can be used as, reductionist models of systems under observation. To make empirical sense, both General Relativity and quantum mechanics require observers and apparatus as given entities, and both are really emergent concepts.

This may not make immediate sense to you now, if you have not studied calculus yet, but the reason I hold my views about General Relativity and quantum mechanics is that each branch of mathematical physics uses *differential equations* in their standard formulations. Differential equations give a localized view of dynamics and look like reductionist physics. Unfortunately, differential equations have to be integrated, or summed up, in order to be useful, and those integrals live on the emergent side of the theoretical fence. You have to put in boundary or initial conditions if you want to get a unique solution to a differential equation. It is a fact that in General Relativity the global topology (large scale structure) of spacetime has to be put in "by hand", and that in quantum mechanics all predictions are based in terms of probability (chance, randomness). Unless you have rather peculiar views about spacetime and probability, each of those concepts are emergent and require observers to make sense of them.

**The Whole is Greater than its Parts**

Many years ago, when personal computers were relatively open to tinkering with, I wrote one or two pretty basic computer games. In those days, a single person could think out a game, write the programme code, and end up with something that was playable.

I remember in particular a game I wrote that simulated the rise and fall of the Roman Empire. When I bought my computer, I had acquired a manual giving details of how to write text on the screen, how to draw lines and colour them in, and so on. That manual was essentially a set of reductionist rules. Nowhere in that book was there any reference to any final program that someone could write.

As I developed my Roman Empire program during my spare time, as a matter of entertainment (even mathematical physicist need rest and diversion regularly), I watched my program grow from a few lines that crashed immediately, to an all-singing and dancing finalized entity that had a life of its own. It just worked. I felt as if I had created a living organism.

But that final, working program was nowhere in the manual, yet every part of that simulation consisted of instructions that *were* in that manual. That is the essential difference between reductionism and emergence. My computer simulation could always be broken down into small, logical steps. But the totality, the whole, was beyond anything specific in that manual.

You may understand from this anecdote why I believe that reductionism and emergence are each fundamental yet disjoint aspects of the Universe. They coexist peacefully. You cannot understand complexity from looking at it in reductionist terms. And you certainly cannot say that reductionist principles will give you an inevitable emergent outcome.

In this respect, I think the following quote from a Nobel Prize winning mathematical physicist really makes the point I driving at:

*The world is full of things for which one's understanding, i.e., one's ability to predict what will happen in an experiment, is degraded by taking the*

*system apart, including most delightfully the standard model of elementary particles itself. I myself have come to suspect most of the important outstanding problems in physics are emergent in nature, including particularly quantum gravity.*
*...those students who stay in physics long enough to seriously confront the experimental record eventually come to understand that the reductionist idea is wrong a great deal of the time, and perhaps always.*

R. B. Laughlin [Laughlin (1999)]

## 24.2 The Theory of Everything and the Death of Physics

Mathematical physics has had a string of technical successes involving the theme of unified field theory. Maxwell showed in 1875 that electricity and magnetism can be understood as manifestations of a single unified field known as the electromagnetic field. Many decades later, mathematical physicists managed to add other phenomena to an ever more complicated unified field structure. First came the weak interactions and then the strong interactions.

The process currently has come to a strange point, sometimes referred to as the *Death of Physics*. For quite a few years now, the *Standard Model* of particle physics has been relatively unchanged. As experimentalists increased the collision energies in their accelerators, the *Standard Model* kept on matching the data *quite well.* It was not perfect, but it was and remains quite good.

The problem was that the data was not revealing any significant new features, almost as if physics had run out of new phenomena to discover. The great technical triumph in 2012 of confirming the existence of the Higgs boson changed nothing in this respect, because after all, the Higgs boson, as an integral feature of the *Standard Model*, had been predicted decades earlier. It would have been more exciting indeed if the Higgs had *not* been detected, because then there would have been more reason to explore outside the standard paradigms of particle theory.

At this time (2019), there is a significant theoretical concept that is being intensively investigated experimentally, and that is *supersymmetry*. Briefly, supersymmetry doubles up all known particle species. The photon has a conjectured partner known as the *photino*, the electron has a conjectured partner known as the *selectron*, and so on. Mathematical physicists would dearly love to have empirical evidence for supersymmetry, because several major problems would get solved in particle theory. *String Theory* is based on supersymmetry for one thing. Theorists have been speculating about supersymmetry for nearly fifty years now, and so far, there has been not a shred of empirical evidence for it. That lack of empirical evidence has contributed significantly to the *Death of Physics* situation that physicists and mathematical physicists find themselves in currently.

## 24.3 Renormalization and the constants of physics

One of the big problems for any mathematical physicist researching on any of the above divides is that mathematical physics has had all its successes with reductionist-level mathematics. By that, I mean using standard principles involving Lagrangians, Hamiltonians, manifolds, and such like. Attempts to cross from reductionism to emergence have been fraught with mathematical ambiguities and bad mathematical behaviour, such as the appearance of infinity in the answers to various calculations.

The problem is, experimentalists don't measure infinities. Either their detectors measure finite signals, or else those detectors are swamped and can give no detailed information. All infinities are only in the minds of mathematical theorists.

Mathematical physicists would very much like to emulate Maxwell's theoretical triumph, the prediction that electromagnetic waves move at the observed speed of light. Theorists would like to predict why the electron has the mass that it has in free space, compared to the mass of the proton and other particles. In practice, theorists start off assuming that unobserved electrons and other fundamental particles have unspecified masses known as *bare masses*. When experimentalists observe those particles, the masses that are observed are not the bare masses but so-called *renormalized masses*. These are built up from the bare masses by the self-interactions of the particle-fields involved (or so the conventional wisdom has it). Theorists would like to understand the relationship between bare and renormalized masses much better. Current techniques of dealing with this and other issues involve assuming spacetime has dimensions *close* to four, a technique known as *dimensional regularization*.

Many mathematical physicists are satisfied with these methods, which are generally empirically successful in certain ways. Others remain concerned. My personal view is that complacency in mathematical physics is always a mistake. Currently, important issues remain, such as explaining particle masses and various constants of physics, such as Planck's constant in quantum mechanics or Newton's constant of Gravitation. No one is sure really whether those constants could ever be "explained". It's possible the view we have of physics is as misguided as the view that the ancient Gauls had of the sky, when they worried about the possibility of it falling down.

The infinities that plague modern quantum field theory have everything to do with the three conceptual divides discussed in the previous section. Observers and their apparatus exist on the classical side of the Heisenberg cut, they are emergent phenomena, and they are obviously endophysical processes. The mathematical physics that has been developed so far has been virtually all about the other side of that fence. If you become a mathematical physicist and decide to work on the observational side, you will face severe challenges. First, traditional mathematics dealing with complexity is currently quite under-developed. Second, conventionally minded mathematical physicists will probably not be sympathetic to your research, *unless* you create some spectacularly successful theory. Then they will flock to your door and acclaim you with prizes.

But if you don't get anywhere quickly, be prepared for a great deal of lack of sympathy. The hard fact is, emergence and complexity are enormous conceptual, mathematical, and theoretical challenges that most people run away from as a matter of policy. It may be hundreds of years, for instance, before any headway is made into any great mathematical understanding of complexity. It's quite possible that the only way complexity will ever be "understood" is computational, using supercomputers employing neural network-type generated programs that no human can possibly "understand" in any traditional way. That would represent a defeat for the 'lone genius' approach to mathematical physics that I have discussed elsewhere in this book.

## 24.4 Dark matter and dark energy

The last two decades or so have seen a revolution in the application of General Relativity to cosmology. Increasingly sophisticated experiments have generated vast amounts of data that can be fitted by General Relativity if certain assumptions are made. These suggest that the entire mass and energy in the Universe consists of a relatively small contribution from "normal" mass and energy, the rest consisting of so-called dark matter and dark energy. Currently, no one quite knows what these dark elements are.

I've said it before in this book but it's worth saying it again. If I were starting out in mathematical physics right now, I would go into dark matter and dark energy research. It's currently a hot topic on the experimental front, and may well transform our theoretical understanding of the Universe. Working on this subject will require knowing all about theoretical and observational cosmology, General Relativity, quantum mechanics, and the Big Bang. That should keep you busy for life.

## 24.5 Concluding remarks

If you do decide to become a mathematical physicist, here is a summary of the points I've tried to get across in this book.

### *Don't be put off*

You can become a mathematical physicist if you take the right steps, regardless of your background. Don't be discouraged by people around you who thoughtlessly tell you such an ambition is a pipe dream. What do they know?

### *Plan*

If you want to become a mathematical physicist, you will have to think in terms of years: school, College, University first degree, University doctorate, post-doctoral researcher, and finally, established mathematical physicist in one or other branch of the many-branched Mathematical Physics Tree.

### *Be kind on yourself*

Doing mathematical physics is not easy. It requires developing your mathematical skills and your physical intuition. All of that will take time. Don't get discouraged when the going gets tough, particularly on the mathematics side. Just perservere. A mathematical physicist does not have to develop new mathematics. Einstein would have been sunk if his friend Marcel Grossmann had not taught him all about the tensor calculus that is needed for General Relativity. But then, Grossmann did not have Einstein's physical intuition.

When you struggle to learn a new branch of maths, a situation I have frequently found myself in, remember to take it easy and *be kind to yourself.* It's part and parcel of conditioning that the reconditioning process, where you learn new things, takes time. The rule that works is *modularize.* Take it step by step, bit by bit. If you give yourself enough time, you will crack most problems.

### *Develop a thick but sensitive skin*

Here's where you have to understand people. You will get your critics. Some of them may be thoughtless and unpleasant. You have to recognize such people for what they are and learn to ignore them. Everybody gets such critics. Even Einstein got them.

However, there will be excellent people who will criticize not *you* but your theory. That's good, that's proper science and debate. Listen to such people and try to understand who is mistaken: you or them. As long as criticism is not personal, then it's all part of mathematical physics. Remember, everybody makes mistakes sooner or later.

### *Mathematical Physics is just another choice*

We're all humans. Keep in mind that if you go into mathematical physics, it's just another way of living your life. Don't get overconfident or smug about being a theorist. If you've been trained to do it, then it's no different than any other profession. You should think of being a mathematical physicist as no better or worse than being anything else.

My last anecdote is given with this in mind. It is at my expense. I learnt that sometimes, what we think of ourselves doesn't necessarily get across to other people. I'm reminded of the tradition in ancient Rome, when a successful general had a Triumph: he rode in a chariot through the streets of Rome, cheered by the crowd and welcomed as a hero. Standing next to him on the chariot would be slave, constantly whispering in his ear *memento te mortalem* (remember you are mortal) [Beard (2009)].

**My Triumph**

When we leave school, we go out into the wider world and on to greater things, hopefully. If things go well for us, it's possible that we start to imagine that we are something special. Perhaps we would like to go back to our old school and show our former teachers just how well we have done for ourselves. It's only natural, isn't it?

So it was with me. I had left school having done sciences at Advanced Level. In particular, I had done Physics, which has always been an interest of mine. I've recounted in earlier chapters some of the things that happened to me at the University of Edinburgh, where I had gone to do a first degree in Astrophysics, and had changed after the first year to Mathematical Physics.

Everything went very well after I changed into Mathematical Physics. I eventually graduated and managed to start a doctorate in *elementary particle physics* at the University of Cambridge. Mark those words well: *elementary particle physics*, as they play a particular role in this anecdote.

One fine day, there I was, some five years after leaving school, coming home from Cambridge for a visit, so full of myself. I was now something, was I not? I was doing a fabulously exotic subject, dealing with quarks and subnuclear structure and all that. Elementary particle physics is what's done at the large Hadron Collider (although that was built much later). I wanted to show off.

I decided to go back to my old school for an unannounced visit.

My luck was in. I went to find *Jock*, as we students called him behind his back. He was my old physics master, a dour Scot who had inspired me in physics.

Fortunately, Jock was still around and was able to see me during a free period. I told him that I was doing a doctorate in elementary *particle* physics. He seemed pleased, but I noticed that his attention was focused more on the current class of students doing Advanced Level Physics than on the success story of a former pupil (me). In particular, he rambled on a bit about one particular pupil in his class who was so good at physics. *Fair enough,* I thought, *Five years ago, he might have been rambling on about me to some visitor.*

After our conversation, he kindly escorted me to the main door of the school on my way out. As luck would have it, who should come by but another master who had taught me those five years before. *Jock* signalled to him, and proudly announced to the man:

"John! Look who's dropped in to see us. It's George! He's doing a doctorate in elementary, **practical** physics."

It was this incident that led me to formulate what I call the *Second Law of Time*: *You can never go back*[1].

# Chapter 25

# Glossary

**classical mechanics**
The theory of motion and interaction of particles and other bodies following the principles established by Galileo, Newton, Lagrange, and many others. Generally successful for 99.999% of household situations. The rest require quantum mechanics.

**boondoggle**
An ongoing process where a large group of people (such as an army, country, company, or group of intellectuals) is engaged in some great enterprise, but it's going wrong for two reasons. First, the leadership/management seem to have no idea or concern that the great enterprise is going wrong. Second, some of the underlings (people lower down the management pyramid) can see that the great enterprise is going wrong and they can see *why* it's going wrong. Unfortunately, there's nothing that they can do about it. Having seen the dangers, the underlings have tried to warn the higher-ups but their message cannot register. The leadership/management is so full of confidence in itself that it cannot take advice from anyone lower down the pyramid.

**Copenhagen interpretation**
A view of quantum mechanics that asserts that systems under observation do not "have" definite properties before observations on those systems are made. The outcomes of observations are contextual, meaning they depend on how the states of systems are prepared and how they are observed.

**critical prediction**
Any reasonable theory will be consistent with standard empirical data, such as predicting attractive rather than repulsive gravitation. This raises the problem of deciding which theory is the one to go for when there's a choice.

One way is to find some prediction of one of the theories that none of the others predict. If that critical prediction is then verified by experiments, then it's reasonable to reject the other theories. A good example is Einstein's theory of General Relativity, which predicts the precession of the planet Mercury's orbit accurately, whilst competitor theories do not.

**displacement activity**
Any time-wasting activity that gives you an excuse for not doing something more useful.

**doctorate**
Same as **PhD**. A three or four year programme of research conducted by a **research student** after their first degree. The objective is to write a **thesis** on a specific theoretical topic. The thesis is examined by experts and, if passed, the student is awarded the academic title of **Doctor**.

**emergence**
The idea that **reductionism** cannot explain everything.

**endophysics**
Physics described from the perspective of an observer considered as part of the Universe and subject to its laws.

**exophysics**
Physics described from the perspective an observer looking in on whatever is being observed.

**Heisenberg cut**
The conceptual divide between the classical world of the observer and the quantum world of the systems being observed.

**mathematical physics**
Theoretical research into the laws of the Universe using any required branches of mathematics.

**peaceful co-existence**
The notion that General Relativity and quantum mechanics describe different aspects of the Universe and that there are no conflicting predictions.

**personal tutor**
In the UK University system, a personal tutor is assigned a group of typically five or six incoming students known as *tutees*, looking after their interests and holding weekly problem solving classes known as tutorials.

**phenomenology**
The high energy particle theorists' buzz-word for the art of fitting particle scattering data to graphical curve predictions generated by ad-hoc and relatively specific models.

**postdoc**
Short for *post-doctoral worker.* A researcher on a one or two-year contract often employed on working on the theory that they studied during their doctorate. Postdocs tend to come from all round the world and frequently do two or more such contracts before landing a permanent academic position.

**postgrad**
Short for *post-graduate student.* A student, already with a first degree, currently studying and researching on a specialist topic under one or more supervisors and aiming to get a doctorate (PhD). A gateway into a career in mathematical physics.

**quantum mechanics**
An empirically successful theory that predicts the behaviour of those systems for which classical mechanical principles fail. Quantum principles are radically different to classical principles and are still being explored.

**realism**
The belief that objects such as particles "exist" and have absolute properties, independently of any observer or observation. Likewise, space and time "exist" independently of observation, according to the realist doctrine.

**reductionism**
The idea that the Universe can be explained by simple rules.

**research student**
Graduate student studying for a doctorate.

**Standard Model**
A reasonably successful theoretical model of particle interactions that unites electromagnetism, weak interactions, and strong interactions. Currently (2019) the model involves 19 experimentally determined parameters.

**thesis**
Same as *dissertation.* A full scale book written over three or more years that is submitted by a doctoral candidate and examined by experts. If it is

good and the doctoral viva successful, then the candidate is awarded their doctorate.

**typo**

Typographical error. These are the viruses of literature, plaguing both word text and mathematical equations. Some reprinted technical books will have an inserted page or even pages of typos spotted by readers. Every author gets them and the only cure is vigilance.

**viva**

A live action examination where examiners ask a candidate direct questions face to face and make their judgement on the candidate's spoken replies.

# Chapter 26

# Notes

### *Chapter 1*

1. Meaning that it can be applied outside of science to many areas of human life.

2. I don't believe the Universe is hostile. I think the Universe has no human values one way or the other. If pressed, the only quality I would attribute to whatever is "out there" is *complete indifference.*

3. I don't imagine there are special neurones for religious beliefs, but I once read that there may be regions of the brain where large scale patterns occur that are associated with such beliefs [Carter (1998)]. Damage those regions and a person may lose their faith, or perhaps start to believe in God.

### *Chapter 2*

1. Alexander is not regarded as "Great" or a good role model in some countries that he invaded.

2. You will notice I capitalize General Relativity but not quantum mechanics. Somehow that seems the right thing to do, despite both scientific frameworks being of equal importance in modern science.

3. I've always understood the difference between *perfection* and *excellence.* You can never be perfect, but you can achieve excellence.

4. It's true that in English, there are some careers that used to have a gendered description, such as *actress* or *authoress*, but those terms have

gone out of fashion by this date. In any case, they referred to the gender of the person doing the job, not the job itself.

### *Chapter 5*

1. Some people think you should also expect to find bankers in police stations (behind bars), but that may be a bit unkind.

2. Precession means the orbit is like a gradually rotating ellipse.

### *Chapter 6*

1. I call this anecdote "An old man's game", when I would prefer to call it "An old person's game". I promised in the Preface to be as accurate as I could in my anecdotes, so I've written what I heard.

### *Chapter 8*

1. Religious leaders have always understood and exploited the fact that regular religious observance is an effective conditioning process.

### *Chapter 11*

1. The average human brain has eighty six billion neurons [Herculano-Houzel (2009)].

2. In fact, what was observed was that light from distant galaxies is *red-shifted* in wavelength. That was then *interpreted* as due to the expansion of the Universe.

### *Chapter 14*

1. It's not the element 'lead' but is so called in English because people once thought it was.

2. Usually, what's written on Sheldon and Leonard's whiteboards in *The Big Bang Theory* is clearly simplistic nonsense, but it looks impressive. Real calculation usually require much larger surface areas to complete.

### *Chapter 16*

1. I'm not sure whether I'm being *ironic* or *sarcastic* here.

### *Chapter 17*

1. Given the near inevitability of disappointment in job-seeking, I came to the conclusion long ago that the best policy was to treat it as a game. You may lose a lot of the time, but you really only need to win one match.

### *Chapter 21*

1. *Thank you, Microsoft*, for dropping the ".awd" format when you got us to upgrade our software to *Windows XP*.

2. portable document format

3. It's generally accepted these days that Newton and Leibniz developed their approaches to calculus independently.

### *Chapter 22*

1. Of course I am envious. I could do with the money.

2. *Many Worlds* paradigms are realist models because they invariably assert that a universal wavefunction "exists".

3. This superposition greatly worried Einstein and led Schrödinger to formulate the famous/notorious *Schrödinger's cat scenario*. In that thought experiment version of the $AB$ experiment, the particle becomes the cat, the state of the cat at $A$ is *alive*, and the state of the cat at $B$ is *dead*. You can appreciate why people have become fixated on this scenario, given that intuitively, we always believe cats are either alive or else dead.

### *Chapter 23*

1. In the case of Ramanujan's work, I have to say that it is so deep that even though I understand all the symbols he used, I don't understand how

he strung them all together to get his results. No one does really, because his astonishing mathematics just came out of him like water out of an overflowing bath. It's people like Ramanujan and Einstein that prove to me that our species is not as stupid as it usually appears.

2. Pythagoras was so weird in his beliefs that I think of him as an ancient, actual version of Sheldon Cooper.

### *Chapter 24*

1. My *First Law of Time* is: *Everything changes.*

# Bibliography

Afshar, S. S. (2005). Violation of the principle of complementarity, and its implications, in *The Nature of Light: What Is a Photon?*, no. 5866 in Proceedings of SPIE, pp. 229–244.

Alpha, M. (2374). Remata'Klan, https://memory-alpha.fandom.com/wiki/Remata'Klan/.

Amabile, T. M. (1983). Motivation and Creativity: Effects of Motivational Orientation on Creative Writers, in *81st Annual Convention of the American Psychological Association* (Department of Psychology, Brandeis University (USA)), pp. 1–23.

Aristophanes (424 B.C.E.). The Knights.

Asimov, I. (1957). *Profession*.

Ayas, N. T., Malhotra, A., and Parthsarathy, S. (2013). To Sleep, or Not to Sleep, That is the Question, *Critical Care Medicine* **41**, 7.

Bacon, F. (1893). *The Advancement of Learning* (Cassel & Company, Limited).

Beard, M. (2009) *The Roman Triumph* (Harvard University Press).

Boyd, R. (2008). Do People Only Use 10 Percent of Their Brains? *Scientific American* https://www.scientificamerican.com/article/do-people-only-use-10-percent-of-their-brains/.

Bulkin, B. J. (2019). *Solving Chemistry: A Scientists's Journey* (Whitefox Publishing Ltd.).

Burridge, H. C. and Linden, P. F. (2016). Questioning the Mpemba effect: hot water does not cool more quickly than cold, *Scientific Reports* **6**, p. 37665, https://www.nature.com/articles/srep37665.

Cannell, D. M. (2001). *George Green: Mathematician and Physicist 1793-1841: The Background to His Life and Work*, 2nd edn. (Society for Industrial and Applied Mathematics (SIAM)).

Carter, R. (1998). *Mapping the Mind* (Weidenfeld and Nicolson (London)).

Colosi, D. and Rovelli, C. (2009). What is a particle? *Classical and Quantum Gravity* **26**, p. 025002 (22pp).

Cook, W. (2010). The joy of uncut pages, http://www.commercenewstoday.com/archives/3916-The-joy-of-uncut-pages.html.

Dawes, G. (2012). Nullius in verba, Nihil in verbis, Sapere aude, *Early Modern Experimental Philosophy: University of Otago* https://blogs.otago.ac.nz/emxphi/2012/02/nullius-verba-nihil-verbis-sapere-aude/.

della Mirandola, G. P. (1496). *Oration on the Dignity of Man* (Henry Regnery Company, Chicago), Translated by A. Robert Caponigri.

Eide, F. (2014). Dyslexia: Was Albert Einstein dyslexic? https://www.quora.com/Dyslexia-Was-Albert-Einstein-dyslexic.

Einstein, A. (1905). Zur Elektrodynamik bewgter Körper, *Annalen der Physik* **17**, pp. 891–921, *On the Electrodynamics of Moving Bodies*, translation in *The Principle of Relativity*, Dover Publications, Inc.

Einstein, A., Podolsky, B., and Rosen, N. (1935). Can quantum-mechanical description of physical reality be considered complete? *Phys. Rev.* **47**, pp. 777–780.

Eno, R. (ed.) (2015). *Analects of Confucius.*

Feynman, R. P. (1985). *Surely You're Joking, Mr Feynman* (Unwin Paperbacks).

Fischbeck, H. J. and Fischbeck, K. H. (1982). *Formulas, Facts and Constants* (Springer-Verlag).

Frank, P. (1947). *Einstein: His Life and Times* (Alfred A. Knopf: New York), Translation from German Manuscript by G. Rosen.

Galilei, G. (1623). *The Assayer.*

Garey, J. (2019). Teens and Sleep: The Cost of Sleep Deprivation, *Child Mind Institute* https://childmind.org/article/happens-teenagers-dont-get-enough-sleep/.

Geneen, H. (1985). *Managing* (Avon).

Giang, V. (2015). What It Takes To Change Your Brain's Patterns After Age 25, *FastCompany.com* https://www.fastcompany.com/3045424/what-it-takes-to-change-your-brains-patterns-after-age-25.

Gödel, K. (1949). An example of a new type of cosmological solutions of Einstein's field equations of gravity, *Rev. Mod. Phys.* **21**, 3, pp. 447–450.

Halligan, P. and Oakley, D. (2000). Greatest Myth of All, *New Scientist, 18 November*, pp. 34–39.

Hauer, R. (2008). *All Those Moments: Stories of Heroes, Villains, Replicants and Blade Runners*, reprint edition (2006) edn. (HarperPaperbacks).

Head, T. (2015). Six interesting musical facts about Albert Einstein, https://www.cmuse.org/interesting-musical-facts-about-albert-einstein/.

Heathers, J. and Nickle, J. (2019). What to do when you don't like vegetables, https://www.precisionnutrition.com/dont-like-vegetables.

Heisenberg, W. (1930). *The Physical Principles of the Quantum Theory*, Dover Edition, 1949 edn. (University of Chicago Press).

Heisenberg, W. (1952). Questions of Principle in Modern Physics, in *Philosophic Problems in Nuclear Science* (Faber and Faber, London), pp. 41–52.

Herculano-Houzel, S. (2009). The human brain in numbers: a linearly scaled-up primate brain, *Frontiers in Human Neuroscience*, pp. 1–11, doi:0.3389/neuro.09.031.2009.

Hitchens, C. (2008). *God Is Not Great: How Religion Poisons Everything* (Atlantic Books (London)).

Hughes, V. (2014). The Tragic Story of How Einstein's Brain Was Stolen and Wasn't Even Special, *National Geographic* https://www.nationalgeographic.com/science/phenomena/2014/04/21/the-tragic-story-of-how-einsteins-brain-was-stolen-and-wasnt-even-special/.

Isaacson, W. (2007). *Einstein His Life and Universe* (Pocket Books).

Itzykson, C. and Zuber, J.-B. (2005). *Quantum Field Theory* (Dover Publications, Inc.), originally published by McGraw-Hill, Inc. in 1980.

Kelley, E. C. (1951). *The Workshop Way of Learning* (Harper and Row, New York).

Kennefick, D. (2005). Einstein Versus the Physical Review, *Physics Today* **9**, doi:10.1063/1.2117822, https://doi.org/10.1063/1.2117822.

Laughlin, R. B. (1999). Nobel Lecture: Fractional quantization, *Rev. Mod. Phys.* **71**, 4, pp. 863–874.

Licata, I. (ed.) (2016). *Beyond Peaceful Coexistence* (World Scientific).

LIGO Scientific Collaboration and Virgo Collaboration (2016). Observation of Gravitational Waves from a Binary Black Hole Merger, *Phys. Rev. Lett.* **116**, 6, p. 061102.

M. L. King, J. (1957). Conquering Self-centredness, *Dexter Avenue Baptist Church in Montgomery, Alabama* 11 August.

Mangen, A. and Velay, J.-L. (2010). *Digitizing Literacy: Reflections on the Haptics of Writing.*

Maris, H. J. (2000). On the fission of elementary particles and the evidence for fractional electrons in liquid helium, *Journal of Low Temperature Physics* **120**, pp. 173–204.

Mencken, H. L. (1956). *Minority Report: H. L. Mencken's Notebooks* (The John Hopkins University Press).

Michelson, A. A. and Morley, E. W. (1887). On the relative motion of the earth and the luminiferous ether, *American Journal of Science* **34**, pp. 333–345.

Mill, J. S. (1869). *The Subjugation of Women* (Longmans, Green, Reader, and Dyer (London)).

Mpemba, E. B. and Osborne, D. G. (1969). Cool? *Physics Education* **4**, 3.

Murty, V. P. and Dickerson, K. C. (2016). Motivational Influences on Memory, in J. R. S. Kim and M. Bong (eds.), *Recent Developments in Neuroscience Research on Human Motivation (Advances in Motivation and Achievement)*, Vol. 19 (Emerald Group Publishing Limited), pp. 203–227.

Newton, I. (1687). *The Principia (Philosophiae Naturalis Principia Mathematica)*, new translation by I. B. Cohen and Anne Whitman, University of California Press (1999).

Newton, I. (1736). *The Method of Fluxions and Infinite Series: With Its Application to the Geometry of Curve-lines* (Henry Woodfall).

Oppezzo, M. and Schwarz, D. L. (2014). Give Your Ideas Some Legs: The Positive Effect of Walking on Creative Thinking, *Journal of Experimental Psychology: Learning, Memory, and Cognition* **40**, 4, pp. 1142–1152.

O'Raifeartaigh, C. and Mitton, S. (2018). Interrogating the legend of Einstein's "Biggest Blunder", *Physics in Perspective* **20**, pp. 318–341, doi:10.1007/s00016-018-0228-9, https://arxiv.org/abs/1804.06768.

Orwell, G. (1949). *Nineteen Eighty-Four* (Secker and Warburg).

PACS (1989). Physics and Astronomy Classification Scheme (PACS): Atomic Nuclei, *Z. Phys. A*.

Paul, L. A. (2014). *Transformative Experience* (Oxford University Press).

Pinker, S. (2015). His Brain Measured Up, http://www.xys.org/forum/db/11/240/207.html.

Planck, M. (1949). *Max Planck, Scientific Autobiography and Other Papers* (Williams & Norgate).

Polya, G. (1971). *How To Solve It*, 2nd edn. (Princeton University Press).

Ramanujan, S. (2000). *Collected Papers of Srinivasa Ramanujan* (AMS Chelsea Publishing), G. H. Hardy (Editor).

Robbins, R. *et al.* (2019). Sleep myths: an expert-led study to identify false beliefs about sleep that impinge upon population sleep health practices, *Sleep Health* doi:https://doi.org/10.1016/j.sleh.2019.02.002.

Royal Society (2008). Royal Society statement regarding Professor Michael Reiss, *Royal Society Website* https://royalsociety.org/news/2012/professor-michael-reiss/.

Rubin, V. C. (1989). Weighing the universe: dark matter and missing mass, in J. Cornell (ed.), *Bubbles, voids and bumps in time: the new cosmology* (Cambridge University Press).

Rumsfeld, D. (2002). News Transcript, DoD News Briefing - Secretary Rumsfeld and Gen. Myers, https://archive.defense.gov/Transcripts/Transcript.aspx?TranscriptID=2636.

Schweitzer, A. (1952). *United Nations World* (UN World Incorporated).

Shah, S. (2006). "If you can't join 'em, beat 'em": Julian Schwinger's Conflicts in Physics, *UCLA Historical Journal* **21**, https://escholarship.org/uc/item/70m7r3r8.

Singh, S. (2005). *Fermat's Last Theorem: The Story Of A Riddle That Confounded The World's Greatest Minds For 358 Years* (Harper Perennial).

Smolin, L. (2006). *The Trouble with Physics* (Houghton Mifflin Harcourt).

Strogatz, S. (2015). Einstein's First Proof, *New Yorker* https://www.newyorker.com/tech/annals-of-technology/einsteins-first-proof-pythagorean-theorem.

Swift, J. (1721). *A Letter to a Young Gentleman Lately enter'd into Holy Orders*, 2nd edn. (J. Roberts (London)).

Taleb, N. N. (2010). *The Black Swan: The Impact of the Highly Improbable* (Penguin Books).

Tate, A. (2019). 5 reasons creative geniuses like Einstein, Twain and Zuckerberg had messy desks and why you should too, *canva.com* https://www.canva.com/learn/creative-desks/.

Temperton, J. (2017). 'Now I am become Death, the destroyer of worlds'. The story of Oppenheimer's infamous quote, https://www.wired.co.uk/article/manhattan-project-robert-oppenheimer.

Tesla, N. (1934). Radio Power will Revolutionize, *Modern Mechanix and Inventions*.

The Observer (1987). *The Observer [London], 30 August.*

Wade, N. (1999). Brain may grow new cells daily, *The New York Times* https://www.nytimes.com/1999/10/15/us/brain-may-grow-new-cells-daily.html.

Walker, M. (2017). *Why We Sleep: Unlocking the Power of Sleep and Dreams* (Scribner Book Company).

Walton, A. (2016). Resolving the Paradox of Group Creativity, *Harvard Business News* https://hbr.org/2016/01/resolving-the-paradox-of-group-creativity.

Woit, P. (2006). *Not Even Wrong: The Failure of String Theory and the Search for Unity in Physical Law* (Basic Books).

Xie, L. *et al.* (2013). Sleep Drives Metabolite Clearance from the Adult Brain, *Science* **342**, doi:10.1126/science.1241224.

Yong, E. (2018). A New Theory Linking Sleep and Creativity, *The Atlantic* https://www.theatlantic.com/science/archive/2018/05/sleep-creati vity-theory/560399/.

Young, D. (2015). Charles Darwin's Daily Walks: The mental rewards of exercise, *Psychology Today* https://www.psychologytoday.com/gb/blog/how-think -about-exercise/201501/charles-darwins-daily-walks.

# Index